Urban Stormwater Management in Developing Countries

Urban Stormwater Management in Developing Countries

Jonathan Parkinson and Ole Mark

Publishing
LONDON • SEATTLE

Published by IWA Publishing, Alliance House, 12 Caxton Street, London SW1H 0QS, UK

Telephone: +44 (0) 20 7654 5500; Fax: +44 (0) 20 7654 5555; Email: publications@iwap.co.uk
Web: www.iwapublishing.com

First published 2005; reprinted 2006
© 2005 IWA Publishing

Printed by TJI Digital, Padstow, Cornwall, UK
Index compiled by Indexing Specialists (UK) Ltd, Hove, UK
Typeset by Gray Publishing, Tunbridge Wells, UK

Apart from any fair dealing for the purposes of research or private study, or criticism or review, as permitted under the UK Copyright, Designs and Patents Act (1998), no part of this publication may be reproduced, stored or transmitted in any form or by any means, without the prior permission in writing of the publisher, or, in the case of photographic reproduction, in accordance with the terms of licences issued by the Copyright Licensing Agency in the UK, or in accordance with the terms of licenses issued by the appropriate reproduction rights organization outside the UK. Enquiries concerning reproduction outside the terms stated here should be sent to IWA Publishing at the address printed above.

The publisher makes no representation, express or implied, with regard to the accuracy of the information contained in this book and cannot accept any legal responsibility or liability for errors or omissions that may be made.

Disclaimer
The information provided and the opinions given in this publication are not necessarily those of IWA, and should not be acted upon without independent consideration and professional advice. IWA and the authors will not accept responsibility for any loss or damage suffered by any person acting or refraining from acting upon any material contained in this publication.

British Library Cataloguing in Publication Data
A CIP catalogue record for this book is available from the British Library

Library of Congress Cataloging-in-Publication Data
A catalog record for this book is available from the Library of Congress

ISBN: 1843390574

Contents

Foreword		*ix*
Preface		*xiii*
Acknowledgements		*xv*
About the authors		*xvii*
1.	Urbanisation and urban hydrology	1
	1.1 Urbanisation and its impacts on urban hydrology	1
	1.2 Urban runoff and climatic factors	6
	1.3 Causes, types and physical impacts of urban flooding	8
	1.4 Environmental impacts of urban runoff	10
	1.5 Institutional challenges	13
	1.6 References	16
2.	Impacts of flooding on society	18
	2.1 Social and economic impacts of flooding	18
	2.2 Health impacts related to drainage and flooding	21
	2.3 Vulnerability and livelihood impacts	24
	2.4 Perceptions of and responses to flooding	27
	2.5 Implications on the design of urban stormwater management systems	29
	2.6 References	31
3.	Integrated framework for stormwater management	33
	3.1 The need for an integrated framework	33

vi Contents

	3.2 Key elements of Integrated Water Resource Management (IWRM)	35
	3.3 Integrated planning of urban catchments	39
	3.4 Integrated management of urban infrastructure and services	43
	3.5 Stakeholder participation and partnerships	47
	3.6 References	49
4.	Policies and institutional frameworks	51
	4.1 Policy formulation	51
	4.2 Policies for runoff control	54
	4.3 Policies related to land use	56
	4.4 Institutional frameworks for policy implementation	58
	4.5 Institutional development and organisational strengthening	60
	4.6 References	64
5.	Planning and assessment of improvement options	66
	5.1 What is planning?	66
	5.2 The planning process	68
	5.3 Review of the existing situation	70
	5.4 Evaluation and comparison of alternative solutions	73
	5.5 Information collection and management	76
	5.6 Spatial mapping and physical information requirements	79
	5.7 References	83
6.	Configurations of urban drainage systems	84
	6.1 Major and minor drainage systems	85
	6.2 Separate and combined drainage systems	87
	6.3 Underground and surface drainage systems	88
	6.4 Attenuation of stormwater runoff	93
	6.5 Urban hydrology design considerations	98
	6.6 Capacity of drainage conduits	101
	6.7 References	102
7.	Ecological approaches to urban drainage system design	104
	7.1 Strategies for sustainable urban drainage	104
	7.2 Rainwater reuse	108
	7.3 Infiltration of stormwater	111
	7.4 Swales	112
	7.5 Constructed wetlands	115
	7.6 Practical demonstration projects of sustainable urban drainage	116
	7.7 References	119
8.	Applications of computer models	121
	8.1 Computer modelling for hydrology and hydraulics	121
	8.2 The modelling procedure	125

Contents vii

8.3	Model calibration and validation	128
8.4	Model application	132
8.5	Integrated modelling of the urban drainage system	138
8.6	References	139

9. Operational performance and maintenance — 140
   9.1 Operational sustainability and performance evaluation — 140
   9.2 Operation and Maintenance (O&M) strategies — 143
   9.3 Solid waste and impacts on operational performance — 145
   9.4 Control of solid waste problems — 149
   9.5 References — 154

10. Non-structural flood mitigation strategies — 155
    10.1 Stages of the flood mitigation cycle — 156
    10.2 Flood mitigation through land use controls — 160
    10.3 Flood proofing and building controls — 162
    10.4 Flood response strategies — 164
    10.5 Flood recovery and rehabilitation — 169
    10.6 References — 171

11. Participation and partnerships — 172
    11.1 Forms and potential benefits of participation — 172
    11.2 Participation in planning and design — 174
    11.3 Partnerships in project implementation — 177
    11.4 Participation in operation and maintenance — 182
    11.5 Participation in non-structural flood control strategies — 183
    11.6 References — 186

12. Economics and financing — 188
    12.1 Urban drainage – a public good — 188
    12.2 Municipal budgeting and accountability — 190
    12.3 Demand and willingness to pay — 191
    12.4 Costs of stormwater management — 194
    12.5 Revenue generation and cost recovery — 198
    12.6 References — 200

Annex 1: Recommended reading — 202
Annex 2: List of contributors — 208

Index — 211

Foreword

Managing urban stormwater in developing countries poses huge challenges and the consequences of its neglect are severe. Inadequate drainage causes needless death and disease and loss of homes, property and livelihoods. Poor stormwater management also pollutes the environment and squanders limited freshwater resources. Sound investment in stormwater management can reduce these losses, but only after setting priorities and making difficult choices. These choices are particularly tough in the developing world, where on the one hand, the cost of poor stormwater management is so high, but, on the other hand, resources are scarce, and other demands for them are also compelling.

Thirty years ago, Emil Chanlett wrote about environmental protection with a holistic vision that was rare then, but has become more common with time. He described environmental management in terms of the balance and priority of demands between public health, comfort, convenience, efficiency, amenity and ecosystem protection. Narrow views of environmental management which are either too 'human-centred' or too 'ecosystem-centred' are not only politically foolish, but also fundamentally miss the point about getting the balance right to achieve the best result. 'Getting the balance right' in setting the objectives is hard enough in the resource-rich countries, which can more easily 'afford' environmental concern; but defining the appropriate balance for urban areas in developing countries is a huge challenge.

In addition to balancing the objectives of sustainable urban development and the right level of investment, there are other balances that those involved in stormwater management must strike. Urban planners, engineers, community activists and other practitioners also need to balance the practice of their art, and must consider social, technical, financial and institutional aspects. They must balance capital investment and operational costs and must balance performance against reliability, and understand methods by which they can make estimates of performance.

These challenges are difficult for any aspect of urban development, but they are particularly challenging for urban stormwater for a variety of reasons:

(1) Stormwater management involves relatively rare and unpredictable events, which limit the interest and attention of the public – whilst water supply and basic sanitation are everyday needs, stormwater management is seasonal, and often of public interest only after a significant failure.
(2) Stormwater management is fundamentally a collective public service. Unlike water supply or basic sanitation, there is virtually no hope of obtaining any significant level of investment from either the individual household or the private sector without state intervention.
(3) Stormwater management is inextricably linked to other public services. Good storm drainage is essential for basic sanitation and decent transportation, but drainage systems without good solid waste management cannot work. As well documented in this book, good stormwater management also requires sound land use planning and management.

Yet, despite its complexity, intellectual challenge, and importance in the everyday lives of the poor in developing countries, there is surprisingly little literature on urban stormwater management in developing countries. Nearly 30 years ago, when I first became aware of storm drainage and its links with sanitation in slums, I felt I had no guidance except the verbal wisdom of senior engineers. There were plenty of technical references on hydraulics and storm drainage for the industrialised world, but there was nothing that talked about the problem (and the limited resources) facing the slum-dwellers in Egypt where I was working at the time, and the government agencies trying to address their needs. Since then, the situation has improved a little with the appearance of one or two manuals, a growing representation from developing countries at international conferences and a growing literature in such journals as *Waterlines*, *Urban Water*, and *Environment and Urbanization*.

Urban stormwater management in developing countries is a subject in which balance is essential, and an understanding of the whole is crucial. Nevertheless, the literature on this topic remains largely fragmented, reflecting the piecemeal advances those of us who have worked in the field have been able to make. There has not, until now, been an integrated reference work, for either the sector professional or the concerned generalist, that lays out the various issues and approaches to be balanced in addressing this important topic.

This book is a 'navigator', a guide to the territory which outlines a variety of issues, and points the reader to the experience and documentation of others facing them. It offers, in one place, a starting point on the whole range of issues involved in stormwater management in developing countries. The subject is surprisingly young and fresh, and this book does not offer all the answers. For, although drainage and stormwater management are nearly as old as the first human settlements, the specific challenge of matching resources, environment, institutions and hydrology in developing countries has only begun to be addressed as distinct from the very different problems of the North. There are many books on the technical issues of storm drainage, but there are far fewer places to turn to for a relevant summary of institutional and financial issues involved in stormwater drainage. None of these references also touch upon both public health and environmental concerns.

This book then, is long overdue, and will be of help as a starting point in stormwater management for those who want to improve the quality of life and the environment for people in the cities of developing countries. I have every confidence that the book will prove useful to those facing these challenges; I only wish it had been available to me 25 years ago in Egypt!

Pete Kolsky
Senior Water and Sanitation Specialist
World Bank
Washington, DC

Preface

For many urban dwellers, especially those who live in developing countries, flooding and environmental health problems related to poor drainage are widespread. At the same time, the agencies responsible for the provision of urban drainage infrastructure generally lack sufficient resources to respond effectively to the stormwater problems that they face. In writing this book, we hope to provide urban managers, planners and engineers with information and inspiration to improve their planning, development and implementation of sustainable solutions to these problems.

The main focus of the book is upon 'developing countries', but experiences from many different parts of the World influence the ideas that are presented. In this context, although there are ongoing debates about the implications of the word 'developing', we have chosen to use the word in the belief that virtually all countries are in a continual process of development. Many countries face similar problems and can therefore learn from each other's experiences. Thus, based upon numerous practical examples and case studies, the book provides descriptions of both traditional and contemporary methods to derive solutions to a wide range of urban drainage problems.

Drainage interventions and control of runoff for flood mitigation are generally assumed to be the domain of the civil engineer. The mixed success that has been achieved through the construction of large-scale and expensive drainage systems leads to the conclusion that there is a need for a more integrated approach to urban stormwater management. Therefore, the concept of Integrated Water Resources Management (IWRM) plays an important role in setting the scene and influencing many of the ideas presented in the book.

However, to achieve the benefits of IWRM in practice requires much more than simply adopting new concepts and the development of new management tools. It requires interdisciplinary teams involving engineers, urban planners, economists, environmental scientists and social scientists who need to actively engage with communities through local politicians, community development workers, social activists, and representatives from non-governmental organisations and community based groups. This requires new working practices, but above all it requires a willingness to engage directly at the grass-roots level with local people in order to gain an understanding and appreciation for what it must be like to live in areas that are poorly serviced by drainage infrastructure and those that are prone to flooding – especially in those areas inhabited by poor communities who cannot afford to live anywhere else.

Acknowledgements

Jonathan Parkinson acknowledges the Brazilian Funding Agency for Scientific Development and Technology (*Conselho Nacional de Desenvolvimento Científico e Tecnológico – CNPq*) for financial support during the course of the preparation of this book. He would also like to thank colleagues from the School of Civil Engineering at the Federal University of Goiás (UFG) in Goiânia, Brazil for their encouragement and support.

Ole Mark would like to thank his colleagues at DHI – Water & Environment, who have provided comments and support for the preparation of this book. He would like to express special gratitude to Ms. Susanne Kallemose, Ms. Birgit Gavilan, Sten Lindberg and Jørn Rasmussen.

The authors would like to acknowledge the United Kingdom's Department for International Development (DFID) who provided funding for the literature review, which formed the initial basis for parts of the book. This review was undertaken as part of a research project (EngKARR8168) funded by DFID and managed by Brian Read from the Water and Engineering Development Centre (WEDC) at Loughborough University in collaboration with GHK International.

The authors thank Dr. Jozsef Gayer – Chairman of the Working Group on Technology Exchange, Transfer and Training (TETT) of the IAHR/IAWQ Joint Committee on Urban Drainage – and Dr. Pete Kolsky, Senior Water and Sanitation Engineer at the World Bank for their advice and encouragement from a very early stage and also for their assistance in reviewing various chapters of the book.

The text has also benefited from contributions from the following persons who kindly offered to review various chapters of the book: Professor Richard Ashley (University of Sheffield), Professor Kapil Gupta, (Indian Institute of Technology, Bombay, India), Dr. Roger Few (University of East Anglia, Norwich), Professor David Butler (Imperial College, London), Kevin Tayler (Honorary Professor – School of City and Regional Planning, Cardiff University), Dr. Manfred Schuetze

(IFAK, Magdeburg, Germany), Dr. Nick Devas (University of Birmingham) and Dr. P.B. Anand (University of Bradford).

In addition, we would like to express considerable appreciation to the following persons for their contribution to the book in the form of case studies, photographs and diagrams, as well as offering additional comments when reviewing the parts of the book relevant to their contributions:

Hans Christian Ammentorp (DHI – Water & Environment), Dr. Nguyen Viet Anh (CEETIA, Hanoi, Vietnam), Dr. Slobodan Djordjevic (Exeter University, United Kingdom), Professor Bonifacio Fernández (Pontificia Catholic University, Santiago, Chile), Dr. Birgitte Helwigh, Dr. Jean O. Lacoursière (Kristianstad University, Sweden), Henrik Larsen (DHI – Water & Environment), Roger Null (Kennedy Jenks, San Francisco, USA), Paul Shaffer (CIRIA, London), Professor Carlos Tucci (Federal University of Rio Grande do Sul, Brazil), Chitra Vishwanath (Rainwater Club, Bangalore, India), Dr. Sutat Weesakul (Asian Institute of Technology, Thailand), Dr. Simon Lewin (London School of Hygiene and Tropical Medicine), Dr. Amit Kapur (Yale University, USA), Shaleen Singhal (TERI, New Delhi, India), Shashi Bhattarai (Integrated Consultants Nepal Pvt Ltd), Khatmandu, Nepal), Associate Professor Dr. Aminuddin Ab. Ghani (River Engineering and Urban Drainage Research Centre, Universiti Sains Malaysia), Professor Manfred Ostrowski and Dirk Muschalla (Technical University, Darmstadt, Germany), Dr. Márcio Benedito Baptista and Dr. Nilo de Oliveira Nascimento (Federal University of Minas Gerais, Belo Horizonte, Brazil), Muhammad Yasin (formerly of Youth Commission for Human Rights, Lahore, Pakistan), Roberto Chávez (Knowledge and Learning Program, Transport and Urban Development Department, World Bank Group), Dr. Alphonce Kyessi (University of Dar es Salaam, Tanzania), Dr. Gift Manase (Institute of Water and Sanitation Development, Harare, Zimbabwe), Morshed Monzu (CARE Bangladesh), Rabin Bogati (CARE Nepal), Martin Strauss (SANDEC, Switzerland), David Satterthwaite (IIED, UK), Rod Shaw (WEDC, Loughborough University, UK), and Erika de Castro (Centre for Human Settlements, Vancouver, Canada).

Finally, the authors would like to thank Alan Click and Alan Peterson from IWA Publishing who have been very supportive during the course of the preparation of the book.

About the authors

JONATHAN PARKINSON BEng (Hons.) MSc DIC PhD

Dr. Jonathan Parkinson is a civil and environmental engineer, specialising in the provision of urban infrastructure and services for sanitation, drainage and wastewater management. He has worked extensively in different parts of Asia and Africa and since his initial experience working in India on a research project managed by the London School of Hygiene and Tropical Medicine, he has developed and expanded his interests related to urban drainage in developing countries, with a particular focus of the needs of low-income communities. Through his consulting experience with GHK International, he developed a broad understanding of socio-economic and institutional factors that influence the sustainability of urban infrastructure and services in developing countries. He recently started to work as a consultant for development co-operation in the water sector for "hydrophil", a company based in Vienna, but the bulk of the work involved in preparation of this book was undertaken whilst working as a visiting researcher at the Department of Civil Engineering at the Federal University of Goiás, Brazil.

OLE MARK MSc PhD

Dr. Ole Mark is a civil and environmental engineer, specialising in modelling of urban drainage and sewer systems. He has spent 14 years at DHI Water & Environment working with consulting, research and software development concerning practical problem solving within urban drainage. Previously, he spent three years as Associate Professor at the Asian Institute of Technology in Bangkok, Thailand where he taught and carried out research within planning, design, and implementation of sewer and stormwater drainage systems. His work has traditionally focused on urban flooding and urban water impacts on the receiving waters. Currently, he is Chief Engineer and Head of Innovation within the Urban Water group at DHI Water & Environment.

DEDICATIONS

Jonathan Parkinson would like to thank his wife, Luiza, for all the support and understanding that she offered during the preparation of this book and he would like to dedicate the book to his son, Lucas.

Ole Mark would like to dedicate this book to his wife, Birgitte, for her endless patience and encouragement at all times.

1
Urbanisation and urban hydrology

This chapter provides the context for many of the issues that are discussed in the rest of the book. It gives an introduction to the process of urbanisation and the consequences for urban hydrology and related physical and environmental impacts of flooding. It highlights some of the pertinent issues within the context of urban stormwater management in cities in developing countries, and considers the institutional challenges in relation to land use, especially in cities that have a high proportion of informal developments and illegal settlements.

1.1 URBANISATION AND ITS IMPACTS ON URBAN HYDROLOGY

Urbanisation is one of the most important demographic trends of the twenty-first century. By 2030, it is estimated that the global urban population will reach 4.9 billion, an increase of 2 billion city people from 2000, which equates to a rise from 47% to 60% of the total population in the world (United Nations 2001). There are a number of important considerations in relation to this growth that should be taken into consideration:

(1) The majority of this growth is concentrated in towns and cities in developing countries and transitional economies.

© 2005 IWA Publishing. *Urban Stormwater Management in Developing Countries* by Jonathan Parkinson and Ole Mark. ISBN: 1843390574. Published by IWA Publishing, London, UK.

Figure 1.1 Buildings on floodplains of natural storm drainage channels in Hyderabad, India. (Photo: Kapil Gupta.)

(2) Although larger urban areas encompass an increasing amount of this population, the proportion of people living in mega-cities (urban agglomerations with more than 10 million inhabitants) will remain relatively small. In 2000, only 4.3% of the world's total population lived in mega-cities, whereas the proportion of the world population living in small cities is considerably larger – 28.5% lived in cities of less than 1 million inhabitants (United Nations 2001).

(3) Much of this urban growth is unplanned, with communities and private developers taking advantage of the weak regulatory capacity of local authorities, particularly in areas outside of municipal boundaries. In some cities, the proportion of the population living in these informal settlements can be as high as 20% (UN-Habitat 2003).

In many instances new buildings occupy floodplains and natural drainage pathways (see Figure 1.1) and the problems of stormwater drainage are frequently worsened by downstream flow constrictions caused by unregulated developments. Many cities lack effective storm drainage systems and ill-planned construction closes off natural watercourses. In some cities, urban wetlands are important physical features of the natural environment that provide essential hydrological functions for flood alleviation and maintain river/stream flows during the dry season. Unfortunately, these benefits are generally ignored as cities develop, during which natural watercourses are often destroyed or relined with concrete and wetlands drained to allow for developments.

As cities develop, the provision of urban infrastructure and services changes according to the level of economic development – both in terms of coverage and quality of the service. Table 1.1 summarises various urban environmental issues related to the *brown* (sanitation) and the *green* (environmental) agenda, and highlights the linkages between drainage, sanitation, water resources and solid waste management. As a result of these linkages, overflows from clogged storm drains and sewers during high rainfall are a major cause of flooding in urban areas.

Table 1.1 Urban environmental issues related to the level of economic development (adapted from World Development Report 2003).

	Low	Lower-middle	Upper-middle	High
Sanitation	Low coverage of household sanitation, low ratio of public toilets to residents, open defecation in some neighbourhoods; high risk of diarrhoeal diseases.	Improved coverage of latrines and public toilets, but often poorly maintained; Low coverage of sewers and no treatment of wastewater.	Increased access to improved sanitation, but many residents in slums and informal settlements not served. Only a small proportion of wastewater is treated.	Full coverage of household sanitation and sewerage connections; Improved wastewater treatment systems.
Drainage	Storm drains very inadequate and poorly maintained; frequent flooding, and high prevalence of water-related disease vectors.	Improved drainage in many areas but low-income areas remain unserved.	Better drainage throughout the city but occasional flooding persists. Pollution in urban watercourses remains a problem.	Good drainage results in reduced flooding. Improved control of pollution from urban runoff.
Water resources	Mixed sewage and stormwater runoff to water bodies causes a wide range of pollution problems and environmental health concerns.	Contamination of surface and groundwaters from poorly maintained latrines and untreated wastewater.	Excessive demands on groundwater and continuing pollution problems from effluents discharge.	Improved effluent controls and treatment to reduce pollution. Increasing focus on wastewater reuse.
Solid waste management	Poor collection services; open dumping or burning of mixed wastes; high exposure to disease vectors (rats, flies).	Moderate coverage of collection service, little separation of hazardous waste; illegal dumping and uncontrolled tips.	Better organised collection of domestic wastes but hazardous waste management remains a problem and landfills poorly managed.	Waste reduction, resource recovery. Improved management of hazardous waste, controlled. Landfills or incineration.

The level of economic development also has implications to urban hydrology and stormwater management in other ways. For instance, the increasing use of the car and other forms of road transport results in a significant increase in impervious areas for the road surfaces and areas for parking. In heavily developed cities, roads and other transport-related impervious surfaces can constitute up to 70% of the total impervious urban areas (Wong *et al.* 2000). This trend is particularly apparent

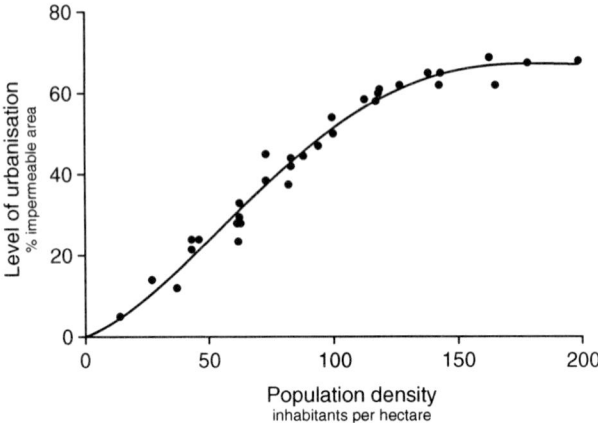

Figure 1.2 Level of urbanisation (% impervious area) plotted against population density in São Paulo, Curitiba and Porto Alegre, Brazil (Campana and Tucci 2001).

in industrialised countries such as the US where the average neighbourhood area per household devoted to streets and parking exceeds the area devoted to housing (Heaney *et al.* 1999). Although there are few other parts of the world where the level of car usage is so prominent, similar trends are observed in many cities due to economic growth and increasing demands for road transportation.

The densification of population living in urban areas and the associated construction of buildings results in dramatic increases in impermeable areas due to paving and roofs. Permanent physical changes to the catchment invariably result in changes to runoff patterns, which affect the magnitude and frequency of flooding. Based on data from Curitiba, Porto Alegre and São Paulo in Brazil, Figure 1.2 illustrates how the percentage of impervious areas increases with the population density.[1] Although other cities follow similar patterns, this relationship should not be used directly to calculate runoff for design purposes in other locations as the curve will invariably be categorised by different development patterns.

In addition to the impermeability of the catchment, the discharge rate and volume of stormwater runoff from urban surfaces depends on other hydrological factors such as the surface depression storage and the antecedent rainfall conditions relating to the wetness of the catchment. The increase in impermeable areas caused by urbanisation has a number of important impacts on the hydrological response from a catchment related to:

(1) Reduced infiltration capacity of catchment surfaces caused by increasing impervious surfaces and compaction of soil, which reduces the capacity of the soil to absorb moisture.

---

[1] This relationship was derived using satellite imagery, and the differentiation between different land use types and catchment surfaces was not made. The remote sensing model was verified with field data showing relative errors below 10% for areas above 2 km^2 and about 20% for areas below 2 km^2 (Campana and Tucci 2001).

Urbanisation and urban hydrology 5

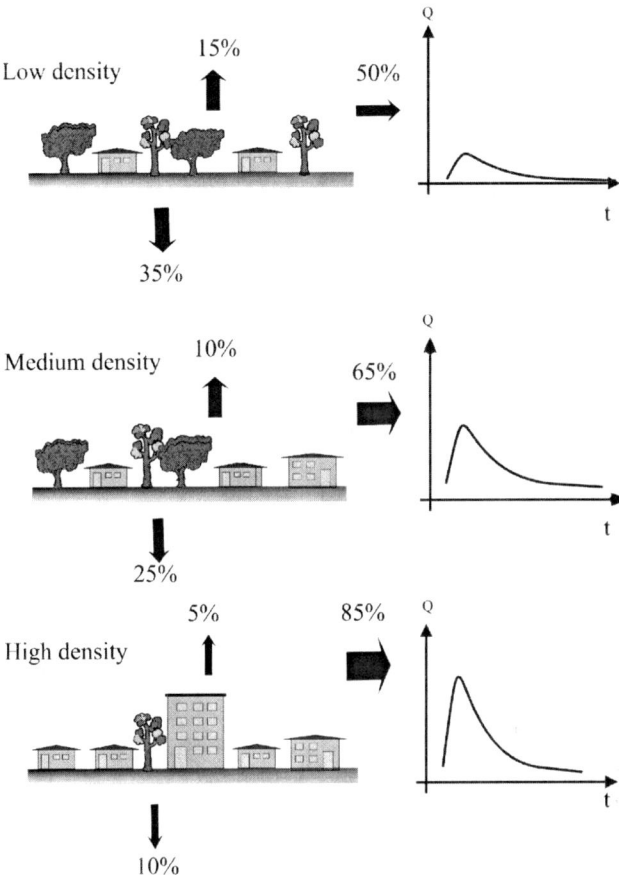

Figure 1.3 Effects of urbanisation on urban stormwater runoff patterns (based on Butler and Davies 2004).

(2) Reduced surface (depression) storage capacity because impervious urban surfaces are much 'smoother' than natural surfaces.
(3) Decreased evapo-transpiration due to the loss in the natural retention capacity of soil, reduced vegetation wetting and interception by plants.

A combination of these factors results in a loss of natural attenuation capacity and runoff from urban catchments is characterised by increases in:

(1) runoff velocity (often measured as time of concentration);[1]
(2) runoff volumes (i.e. the proportion of precipitation that becomes runoff);
(3) Discharge rates and flood peaks.

As shown by the runoff hydrographs in Figure 1.3, the loss of attenuation capacity caused by urbanisation gives flood events a 'flashing' appearance, which often causes

---

[1] The 'time of concentration' is measured by the time it takes rainfall falling on the most remote part of the catchment to reach the outfall.

hydraulic overload of stormwater drainage systems and consequently flooding. However, urban catchments in developing countries often have significantly different physical characteristics from those in industrialised countries, which cause wide variations in the rainfall-runoff response and the resultant volume of runoff and peak flows. Typically, these include lower percentages of impermeable areas and higher depression storage, which means that the runoff volumes and peak discharges are not always as high as would be expected in comparison with catchments of similar population densities in developed countries.

1.2 URBAN RUNOFF AND CLIMATIC FACTORS

In addition to the physical characteristics of the catchment, the other main factor that affects runoff is climate – in particular, rainfall intensities and duration. Many developing countries are located in tropical or subtropical climates where rainfall is characterised by large seasonal variations with a highly pronounced wet season in which the annual rainfall is concentrated during a few months only. The annual rainfall volume and the intensities are generally very high in the humid tropics compared with those in temperate climates. As shown in Figure 1.4, the highest rainfall is in the Andean region, West Africa and South-East Asia.

The rainfall in these regions is generally convective which is characterised by short, but very high rainfall intensities – sometimes exceeding 100 mm hr^{-1} under extreme storm conditions. This type of rainfall is critical for small urban catchments, which have short times of concentration, and has significant influences on the design of urban drainage systems as peak rainfall intensities are the frequent cause of flooding.

Figure 1.5 shows the average peak hourly rainfall intensities recorded in Brasilia, Porto Alegre and São Paulo for each month over a period from 1995 to 2001. These data illustrate the widespread seasonal differences in rainfall in different parts of Brazil and similar variations are observed in other parts of the world. The rainfall hyetograph shown in Figure 1.6 for the rain event in July 2000 resulted in severe

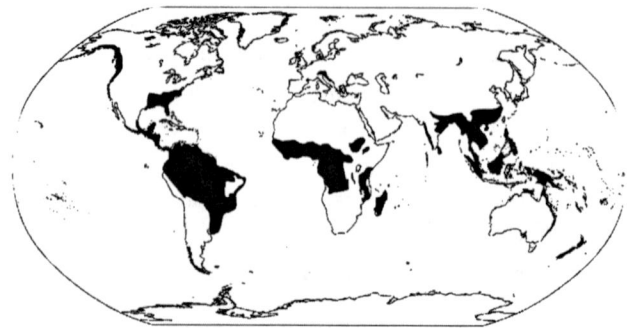

Figure 1.4 Global rainfall distribution showing areas of the world with more than 1.5 m/year average rainfall. (Based on Cairncross and Ouano 1991. Reproduced with permission of WHO using a base map produced by the Nations Online Project.)

flooding, causing widespread disruption, which brought the city of Mumbai to a standstill.

Other climatic factors, such as wind and temperature also affect the scale and nature of the problems related to urban drainage. Urbanisation on a big scale affects the micro-climate, which in return affects the rainfall distribution. Recent scientific studies suggest that climate change will cause shifts in the global rainfall patterns and subsequently increase the intensity of rainfall, and therefore the severity of flooding events. In cases where the global climate change is not taken into consideration, it is likely that there will be an increase in the flood risk for

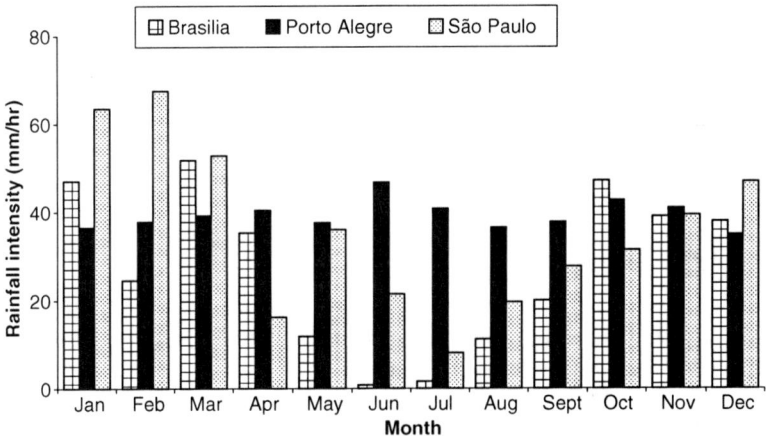

Figure 1.5 Average peak hourly rainfall intensities recorded in Brasilia, Porto Alegre and São Paulo from 1995–2001. (Data reproduced with kind permission of National Institute of Meteorology, Brazil.)

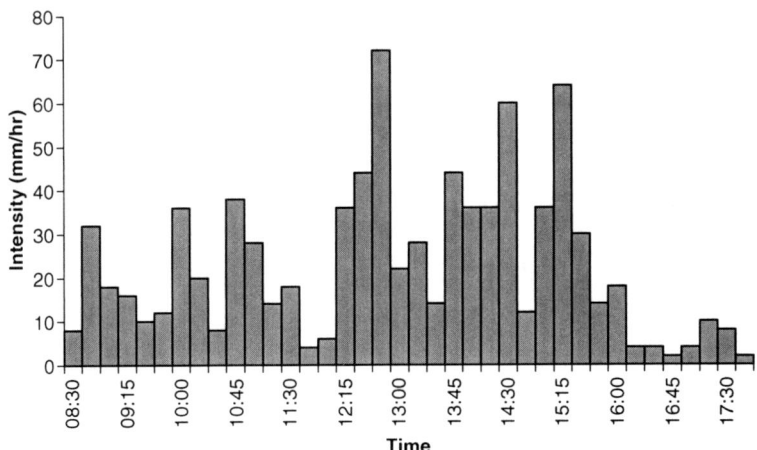

Figure 1.6 Rainfall hyetograph from Santa Cruz in Mumbai, India – 2000 (Gupta 2002).

many human settlements and the impacts will fall disproportionately on the poor (McCarthy *et al.* 2001).

Potential climate changes may be taken into account as part of the planning, design and management of the urban stormwater system, and the first step is to modify design storms accordingly. However, a detailed description of global climate changes, how to deal with them and their impacts is beyond the scope of this book.

1.3 CAUSES, TYPES AND PHYSICAL IMPACTS OF URBAN FLOODING

The expansion of urban areas and the associated increase in impermeable areas, combined with the tropical rainfall conditions described above, are responsible for the increase in the frequency of urban floods. This situation is aggravated by the lack of planning and delays in the construction of drainage infrastructure that are common in cities in developing countries. However, although flooding is often associated with the disastrous consequences of large-scale storm events, there are also frequent minor flood events, resulting in more localised drainage problems caused by a deficiency in drainage infrastructure as shown in Figure 1.7.

Even though these smaller events are generally not considered to be of serious concern compared with large flood events, the problems associated with this type of flooding may be considered to be more of a problem by affected communities (see Chapter 2). The main types and causes of urban flooding are illustrated in Figure 1.8 and, as described in Table 1.2, these flood events are categorised according to the extent of flooding and the resultant impact.

Figure 1.7 Lack of infrastructure results in poor drainage in Warangal, Andhra Pradesh, India. (Photo: Bruce Pollock.)

Urbanisation and urban hydrology

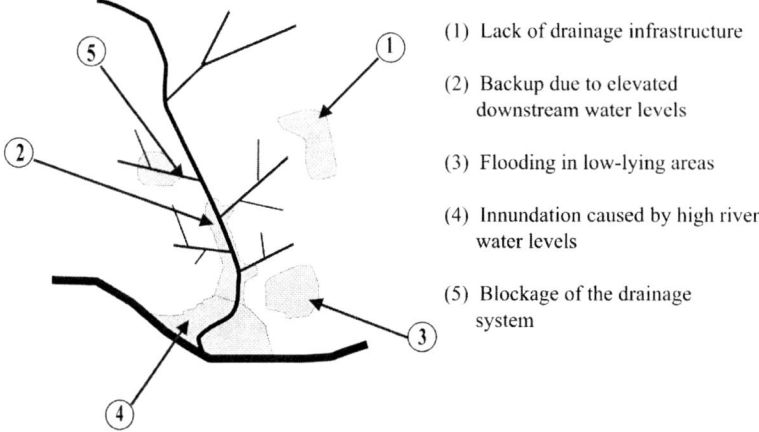

(1) Lack of drainage infrastructure

(2) Backup due to elevated downstream water levels

(3) Flooding in low-lying areas

(4) Innundation caused by high river water levels

(5) Blockage of the drainage system

Figure 1.8 Causes and types of urban flooding.

Table 1.2 Typology of flood types, characteristics and impacts.

Flood type	Characteristics of flooding and impacts
Minor	Localised flooding caused by inadequate drainage of stormwater runoff can occur frequently where drainage infrastructure is lacking or deficient. The main impacts of these events relate to deterioration in environmental health conditions – notably water-related diseases – and gradual structural decay due to dampness and water logging.
Moderate	Flood events of this type occur less frequently than minor floods, but affect larger areas. The impacts may include temporary disruption to transportation systems and inconvenience to city life as well as all of the above problems. These events also cause erosion of sediments leading to pollution in receiving waters, may contribute to the propagation of water-related diseases and can cause structural damage (but not as severe as those related to major events).
Major	Large-scale inundation (approximately 1–2 times per year) may cause widespread disruption and damage affecting communities and businesses. Can cause erosion and instability of soils on steep slopes threatening houses and other buildings in risk areas. A wide range of urban infrastructure and services, such as water and power supply, can be affected. A combination of these effects causes significant economic impacts and recovery from major events can take many days or even weeks.
Extreme	Extreme events occur infrequently (once every few years or less) and result in inundation for a prolonged duration. The impacts are severe to catastrophic and loss of life is likely, especially in areas of steep unstable slopes prone to landslides. These events often reach the international headlines due to the dramatic scale of the impacts and structural damage. Recovery from these events can take a long time and economic repercussions can be devastating for those who do not have resources to recover.

While the terms flooding and inundation are often used interchangeably, *flooding* tends to be related to drainage-related problems where there is insufficient capacity in the drainage system whereas *inundation* refers to the rising of a body of water and its overflowing onto normally dry areas, such as is the case where a river flows over its banks and onto the floodplain.

1.4 ENVIRONMENTAL IMPACTS OF URBAN RUNOFF

As well as the increase in frequency and magnitude of urban flooding, urbanisation results in pollution problems in urban streams and other receiving waters. Much of this is caused by discharge of wastewaters during dry weather conditions, but wet weather has the effect of cleaning urban surfaces and drainage channels, resulting in significant pollution problems. Thus, the quality of runoff is influenced by many factors, including land use, waste disposal and sanitation practices. Figure 1.9 illustrates some of the pollution problems relating to waste discharges into urban drainage channels. In addition, high rainfall intensities have a particularly high-erosion capacity and suspended solids concentrations in runoff can be very high – particularly as a result of construction activities.

A significant amount of pollutants ranging from gross pollutants to particulates and soluble toxins are generated from urban catchments. There are a number of pollutants of principal concern in urban runoff and these affect organisms in receiving waters in various ways (see Table 1.3). It is important to note that these environmental stressors may interact to varying degrees in an antagonistic, additive or synergistic fashion (Porto 2001) – meaning that the cumulative effect of different pollutants is likely to be worse than the sum of individual effects of each one.

Figure 1.9 Typical pollution problems in an urban stream in an informal settlement (favela) in peri-urban São Paulo – Brazil. (Photo: Jonathan Parkinson.)

Table 1.3 Environmental impacts of pollutants from urban stormwater runoff in receiving watercourses.

Pollutant	Source	Environmental impact
Oxygen demanding materials	Vegetation, excreta and other organic matter.	Depletion of dissolved oxygen concentration, which kills aquatic flora and fauna (fish and macro-invertebrates) and changes the composition of the species in the aquatic system. Odours and toxic gases form in anaerobic conditions.
Inorganic compounds of nitrogen and phosphorus	Fertilisers, detergents, vegetation, animal and human urine, sewer overflows and leaks, septic tank discharges.	In high concentrations, ammonia and nitrate are toxic. Nitrification of ammonia micro-organisms consumes dissolved oxygen. Nutrient enrichment (eutrophication) causes excessive weed and algae growth blocks sunlight, which affects photosynthesis and causes oxygen depletion.
Oils, greases and gasoline	Roads, parking areas, garages and petrol stations (spillages and leakages of engine oil), and industry. Vegetable oils from food processing and preparation.	Pollution of drinking water supplies and impacts on recreational use of waters. Reduction of oxygen transfer at the water surface. Carcinogenic compounds may cause tumours and mutations in certain species of fish.
Heavy metals, pesticides, herbicides and hydrocarbons	Industrial and commercial areas. Leachate from landfill sites and improper disposal of household chemicals.	Toxic to aquatic organisms and accumulate in the food chain impairing drinking water sources and human health. Many of these toxins accumulate in the sediments of streams and lakes.
Suspended solids, sediments and dissolved solids.	Erosion from construction sites, exposed soils, street runoff and stream banks.	Sediment particles transport other pollutants that are attached to their surfaces. Sediments interfere with photosynthesis, respiration, growth and reproduction, and deposited sediments reduce the transfer of oxygen into underlying surfaces.
Higher water temperatures	Increase in water temperature as runoff flows over impervious surfaces (asphalt, concrete, etc.).	Reduced capacity of water to store dissolved oxygen. Impact on aquatic species that are sensitive to temperature.
Trash and debris	Domestic and commercial refuse, construction waste and various types of vegetation.	Blockages and constrictions to drainage channels, aesthetic loss and reduction in recreational value.

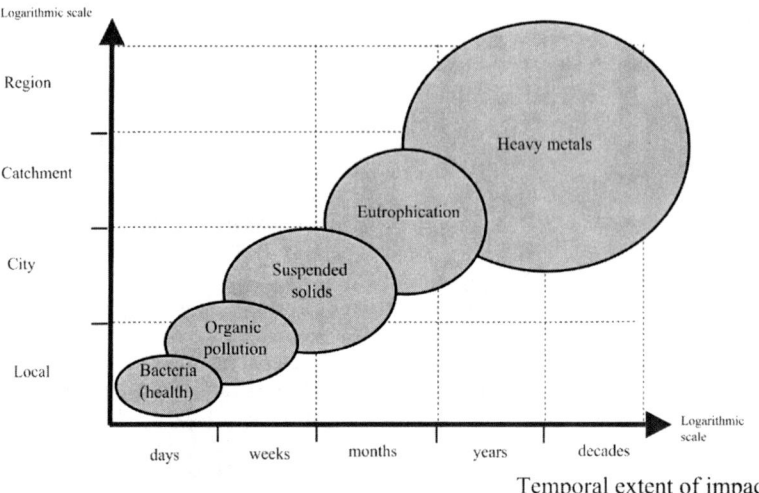

Figure 1.10 Temporal and spatial extent of urban water quality problems.

Runoff from roads and other paved areas is of concern as it harbours a vast array of particulates and chemicals arising from the activities that characterise the land use. According to Wong *et al.* (2000), roads and other transport-related impervious surfaces contribute a higher proportion of stormwater pollutants than other impervious surfaces (e.g. roof areas, pedestrian pathways, etc.). Runoff from transport-related surfaces consistently show elevated concentrations of suspended solids and associated contaminants (such as lead, zinc and copper), as well as other pollutants (such as hydrocarbons). In addition, as described in more detail in Chapter 2, problems related to microbiological pollution are caused by the flooding of sanitation systems and the discharge of pathogenic bacteria and other micro-organisms (viruses, protozoa, etc.) can cause intestinal infections.

Figure 1.10 illustrates how the nature of the pollutant influences the temporal and spatial extent of urban water quality problems. Typically, pollution problems related to discharge of stormwater occur within relatively short duration called *acute impacts*, although there may also be *chronic impacts*, depending on:

(1) The nature of the pollutant, for example impacts related to oxygen demand or toxicity from high ammonia concentrations will be short-term, but nutrient enrichment (eutrophication) will only occur in the long-term.
(2) The nature of the receiving water, the impacts will depend upon the flow conditions – for instance, stream, river or lake. In general, chronic pollution problems from storm runoff are rare in rivers – except where solids from deposited sediments smother the riverbed.

1.5 INSTITUTIONAL CHALLENGES

Many of the flood and pollution-related problems associated with urban runoff described above are common to cities in many different parts of the world (Tucci 2001). However, these problems present a particular challenge in developing countries due to various problems associated with the institutional arrangements for urban drainage. One of the key institutional issues relates to the fact that often drainage has no clear constituency until major problems occur (World Development Report 2003), and it is only after large-scale flood events that investments to improve the infrastructure are made.

One of the main problems in developing countries is that there is insufficient control over new developments due to deficiencies in the administrational systems for urban planning and control. A particular problem relates to the control of informal settlements, which may have a distinct set of drainage problems and a complete lack of infrastructure to drain stormwater. In these situations, buildings are constructed with no consideration for stormwater drainage and where these occupy floodplains or natural drainage pathways, the problems of stormwater drainage are increased due to the restricted flow capacity.

As described in further detail in Chapter 4, institutional problems affect broad areas of operational performance, which are qualitatively different from specific technical or procedural problems (Cullivan *et al.* 1998). Planning authorities and regulatory agencies often lack resources to develop and implement effective solutions for the control of runoff and mitigation of the flood events. Whilst the separation between city and disaster management continues, valuable opportunities for reducing urban risk are lost. Even where national disaster management systems have been formally created, good co-ordination between different government and other organisations does not necessarily exist, leading to confusion, contradictions, overlapping functions and gaps in responsibility (Sanderson 2000).

Other institutional problems relate to a lack of co-ordination between different agencies and organisations with interests in urban drainage. In addition, urban flooding is not bound by local administrative boundaries, because stormwater drainage and protection facilities are part of an environmental system that is larger than an incorporated city territory. The definition of the boundary areas also results in problems associated with the poor alignment between administration and hydrological boundaries (see Chapter 3). Lack of effective urban planning and management in developing countries is a widespread issue affecting urban drainage systems. The example described in Box 1.1 and illustrated in Figure 1.11 presents a particularly important issue related to the ownership of land. In this example, one private landowner managed to halt the construction of a large-scale drainage system in Dhaka for over a year.

The drainage problems described are further aggravated because cities in developing countries grow rapidly, but without the funds necessary to extend and rehabilitate their existing drainage systems. Due to the lack of effective urban planning systems in many developing countries, the control over new developments

> **Box 1.1 Problems of drainage construction in Dhaka**
>
> In 1996, as part of the drainage improvement plan of the Dhaka Metropolitan city, Dhaka Water and Sewer Authority (WASA) initiated a scheme to rehabilitate the natural channel section of the Segunbagicha Khal by replacing it with a concrete box culvert. The total length of the culvert is 2.1 km which varies in cross-section but on average is approximately 4 m by 4 m. After completion of 85% of the total works the construction was stopped by court order due to a land dispute with one private landowner. During the land dispute, the stormwater was drained to a storage pond, and the water flowed along an open channel to the basin in front of the sluice gate. Finally, in the autumn of 2000, the dispute was resolved and, since then, the culvert has been in operation.

Figure 1.11 Construction of large drainage conduit halted by a land dispute in Dhaka, Bangladesh during 1996–2000. (Photo: Ole Mark.)

is often weak and in direct conflict with drainage requirements. For example, the photo in Figure 1.12 shows that developers can operate with complete disregard for planning regulations and, in this situation in Hyderabad, India, the developer is actually constructing a multi-storey building that partially blocks one of the main drainage channels.

Another example, which highlights the institutional issues that are commonly encountered in cities in developing countries is described in Box 1.2. This example highlights some of the institutional problems related to the provision of urban

Figure 1.12   Private developers construct buildings and obstruct a storm drain in Hyderabad, India. (Photo: Kapil Gupta.)

Box 1.2   Institutional issues related to urban catchment planning in Delhi

With the Supreme Court of India passing a jurisdiction on the governing authorities and their development agencies to implement programmes to clean up the rivers of India, the government constituted a task force in July 1993 to review a proposal for covering of open drains in Delhi – in order to reduce the environmental health problems, especially for those with squatter settlements on their banks. The study observed that many of the problems arose due to the lack of foresight and effective urban planning, highlighting the conflict between the lack of open space in the city and the availability of space alongside the drainage channels.

The study also highlighted the institutional problems related to the fact that Delhi is divided into different areas with different institutional responsibility for stormwater management assigned to the Municipal Corporation of Delhi, New Delhi Municipal Committee and the Cantonment Board. In addition, the overall responsibility of the drainage system at the Master Plan and Zonal Plan levels rests with the Delhi Development Authority, but this agency does not involve itself in the areas under the jurisdiction of the other governing bodies. The situation is further complicated by the fact that the Irrigation and Flood Control Department of Delhi is responsible for management of the larger drainage system.

Source: UNESCO (2001)

---

drainage and effective stormwater management systems, and illustrates the institutional complexities that exist in the management of the urban environment.

Box 1.3 describes experiences from Cape Verde, which exemplify the complexity of urban drainage problems in cities in the developing world and point towards some of the solutions that are presented in this book.

> **Box 1.3 Complexities of flood management in Cape Verde**
>
> New dwellers have built houses on the slopes and streambeds of the marginal areas in the town of Praia in Cape Verde. Unlike most rural communities, who have a good understanding of the impact of floods on their livelihoods, the awareness of the problems in urban areas is less evident. Most of the new dwellers responsible for the occupancy of the river beds of the watershed are recent immigrants to the area and they are more interested in resolving their immediate housing problems than those related to flooding. In addition, these residents produce domestic refuse and wastewaters, which are discharged into the local watercourses and diversion canals. In doing so, they pollute the environment and create greater flood risks and environmental health concerns.
>
> In the past, several watershed management and water resources projects have failed to generate expected benefits due to insufficient involvement of beneficiaries during project implementation. There is therefore a need for the development of a more comprehensive approach towards the management of water resources and flooding, requiring the integration of economic development plans and national environmental protection. Special attention should be paid to elements of non-structural flood management with greater emphasis on process-oriented planning and participation of local stakeholders in decision-making. This requires a greater focus on awareness-raising and education amongst the public combined with professional training to enable those responsible for the design of flood control strategies to have a broader understanding of flood-related problems and also a greater appreciation of the role of participation and partnerships in stormwater management. It is also recognised that interagency co-operation in the planning and co-financing is critically important for sustainable water resources development and flood control projects.
>
> *Source*: Sabino *et al.* (1999)

1.6 REFERENCES

Butler, D. and Davies, J. (2004) *Urban Drainage*. E & FN Spon, London.
Campana, N. A. and Tucci, C. E. M. (2001) Predicting floods from urban development scenarios: case study of the Dilúvio Basin, Porto Alegre, Brazil. *Urban Water* 3(1/2), 113–124.
Cairncross, S. and Ouano, E. A. R. (1991) *Surface Water Drainage for Low-income Communities*. WHO/UNEP, World Health Organization, Geneva, Switzerland.
Duncan, H. P. (1999) *Urban Stormwater Quality: A Statistical Overview*. Report 97/1 Cooperative Research Centre for Catchment Hydrology. Monash University, Melbourne, Australia.
Foster, S. S. D. (2001) The interdependence of groundwater and urbanisation in rapidly developing cities. *Urban Water* 3, 185–192.
Gupta, K. (2002). Living With Real Rainfall. *Sewer Systems and Processes Network (SEWNET) Meeting*, 3rd December, Bristol, UK.
Heaney, J. P., Pitt, R., Field, R. and Chi-Yuan, F. (1999) *Innovative Urban Wet-Weather Flow Management Systems*. National Risk Management Research Laboratory Office of

Research and Development. US Environmental Protection Agency. Cincinnati, OH 45268. EPA/600/R-99/029.

McCarthy, J., Canziani, O., Leary, N., Dokken, D. and White, K. (eds) (2001) *Climate Change 2001: Impacts, Adaptation and Vulnerability.* International Panel on Climate Change, Cambridge University Press, Cambridge.

Porto, M. (2001) *Water Quality of Overland Flow in Urban Areas.* Tucci (ed.) UNESCO International Hydrology Programme (IHP-V) Technical Documents in Hydrology **40**(1), UNESCO, Paris, pp. 103–124.

Sabino, A. A., Querido, A. L. and Sousa, M. I. (1999) Flood management in Cape Verde – the case study of Praia. *Urban Water* **1**(2), 161–166.

Sanderson, D. (2000) Cities, disasters and livelihoods. *Environment and Urbanization* **12**(2), 93–102.

Tucci, C. E. M. (2001) *Urban Drainage in the Humid Tropics.* UNESCO International Hydrology Programme (IHP-V) Technical Documents in Hydrology **40**(1), UNESCO, Paris 2001.

UNESCO (2001) *Open Space Planning – To Rid Delhi of its Drainage Dilemmas.* http://www.unesco.org/most/isocarp/proceedings2001/cases/cs01_1453/paper2001.html

UN-Habitat (2003) *The Challenge of the Slums.* Earthscan, London.

United Nations (2001) Population, environment and development in urban settings. Chapter 6 In: *Population, Environment and Development: The Concise Report.* ST/ESA/SER.A/202 United Nations Population Division, Department of Economic and Social Affairs, New York.

Wong, T., Breen, P., Somes, N. and Lloyd, S. (2000) *Water Sensitive Road Design – Design Options for Improving Stormwater Quality of Road Runoff.* Co-operative Research Centre for Catchment Hydrology. Technical Report 00/1. Monash University, Australia.

World Development Report (2003) Sustainable development in a dynamic world. Chapter 6 In: *Getting the Best from Cities.* World Bank, Washington, DC, USA.

2
Impacts of flooding on society

The objective of this chapter is to provide the reader with an understanding of the many ways in which inadequate drainage can affect the health, well-being and livelihoods of communities. Flooding encompasses a wide range of events of varying magnitude and duration, resulting in differing degrees of physical and environmental impacts described in Chapter 1. The impacts of flooding and poor drainage are both complex and interlinked, and affect the livelihoods of communities in a variety of ways. This chapter focuses on how these affect society – both directly and indirectly – and as individuals and collectively. An awareness and understanding of the scale and diversity of these impacts, *who* is affected and how they respond to these impacts can help to identify the most appropriate structural and non-structural interventions within an integrated framework for urban stormwater management.

2.1 SOCIAL AND ECONOMIC IMPACTS OF FLOODING

Flooding causes widespread disruption to city life, essential public services and transport systems (see Figure 2.1). Physical impacts of flooding include damage to property, infrastructure and contents of buildings. Floodwaters can cause failure to pumping stations and damage to water supply systems resulting in a loss of drinking water, as well as other forms of infrastructure and urban services such as electricity supply and waste disposal systems.

© 2005 IWA Publishing. *Urban Stormwater Management in Developing Countries* by Jonathan Parkinson and Ole Mark. ISBN: 1843390574. Published by IWA Publishing, London, UK.

Figure 2.1 Flooding in Dhaka causes widespread disruption to city life. (Photo: DHI – Water & Environment.)

The direct impacts on people are also significant. Large-scale flooding can force the involuntary relocation of families living in high-risk flood areas. Flood events can result in injuries – either directly due to high flows of runoff or, more frequently, indirectly due to accidents caused by the disruption and damage caused by flooding. In addition to the health impacts associated with flooding, there are many perennial health impacts caused by poor drainage of runoff as described in Section 2.2. Incidence of illness and disease often increases after flooding, which can place further demands on medical services, which may also be disrupted by the flooding.

Although flood impact assessments usually focus on the detrimental impacts, there may be positive consequences (e.g. the replenishment of groundwater) to be considered. These benefits may also be seen by communities to be an important part of natural cycle (e.g. for the production of crops) and rainfall may be linked to religious beliefs. Therefore, as described later in this chapter, public perceptions of flooding and drainage problems are highly subjective and affected by many other factors that influence their livelihoods.

2.1.1 Economic impacts

For a minority, there may be other benefits linked to economic opportunities through the provision of services to help those affected by a large-scale flood event, but by far the majority of people are adversely affected by flooding (Few 2003). Economic losses typically affect a much wider area than the area directly affected by flooding and may last much longer than the flood itself. Local and regional economies may be affected by major flood events, which may subsequently have implications on the national economy (WHO 2002a).

The economic impacts of floods are often much greater than those associated directly with the damage caused by the physical impacts of floodwater, such as loss or damage to possessions, ruined food and loss of production during flood-induced power failures. In addition, economic impacts relate to the costs associated with the provision of emergency services, community support and cost of treating the sick. Businesses and factories may be affected too, depending on the type of the product sold and the nature of the service provided. Small businesses are particularly vulnerable from flooding and may face loss of trade and damaged stock as well as leading to a loss of working hours during the time of the flood and during the clean-up (see Figure 2.2).

2.1.2 Psychological effects

The psychological effects of flooding are complex and difficult to quantify in monetary terms. Many of these relate to the uncertainty and confusion before and during evacuation and feeling of personal vulnerability. Other causes of distress are caused by the trauma of the clear-up, living in damp and damaged properties after the flood, lack of practical advice and emotional support as well as increased financial worries (Andjelekov 2001). The World Health Organization recognises that ill health, particularly in the form of psychological problems, may persist for a long time after a significant flood event and these have traditionally been underestimated in the assessment of the consequences of flooding (WHO 2002b). Social support systems may therefore need to be improved to enhance coping strategies related to the stress of flood impacts as well as the health impacts (Tapsell 1999).

Figure 2.2 Small businesses are particularly vulnerable from flooding. (Photo: Ole Mark.)

2.2 HEALTH IMPACTS RELATED TO DRAINAGE AND FLOODING

Although there is increasing recognition of the impacts of large-scale flood events and the damage that these inflict upon affected communities, there is generally less recognition of the prolonged impacts caused by the problems of smaller-scale flooding that affect people living in poorly drained areas. As mentioned in Chapter 1, for many of these communities, it is the regular sickness and the daily inconvenience of living in those areas that are likely to be a greater concern than infrequent large-scale flood events.

Health-related problems associated with poor drainage and flooding are harder to quantity than those caused by physical impacts, but the effects have major implications on the livelihoods of urban dwellers. These health impacts include respiratory illnesses (such as cough, wheeze, asthma and bronchitis) related to persistent dampness and mould in the home caused by poor drainage and a wide range of waterborne diseases.

Runoff from urban surfaces has a very low concentration of pathogens, but the risk of transmission of waterborne diseases increases dramatically when runoff is mixed with wastewater and faecal sludges in foul drains, septic tanks and leach pits, which spreads pathogens around the environment. Therefore, health problems are generally more prominent during the wet season and are linked to a deterioration in sanitation and environmental health conditions related to poor drainage. In addition, there are health impacts during the dry season when open storm drains carrying urban wastewater are potential sources of infection and sites for breeding of vectors.

Details of the potential disease transmission routes related to drainage are described below and illustrated in Figure 2.3.

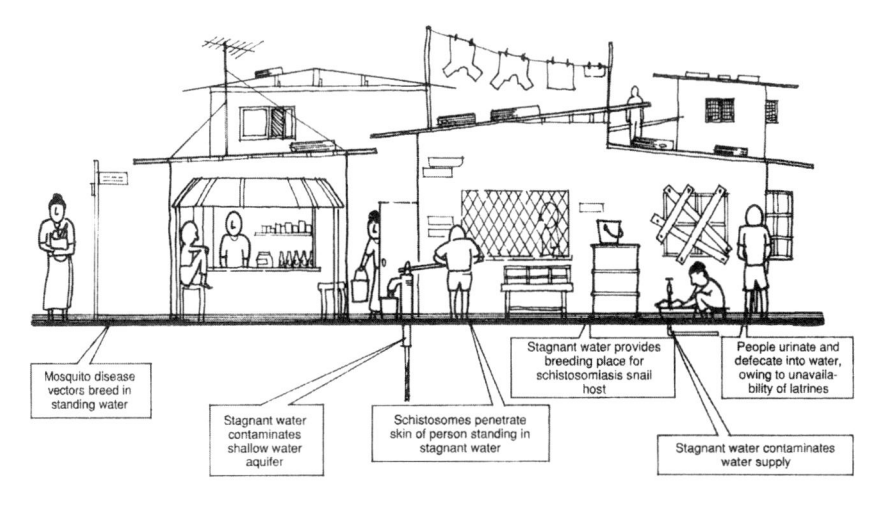

Figure 2.3   Health consequences of poor drainage (Cairncross and Ouano 1991).

2.2.1 Direct contamination and ingestion of pathogens

The most common causes of disease transmission are those related to direct contamination of:

(1) the household environment by intrusion of floodwaters, leading to oral ingestion of enteric pathogens;
(2) water supplies by infiltration of wastewater, especially in systems operating under low-pressure conditions, which may be caused by intermittent operation or power failures due to flooding;
(3) watercourses used for cooking, cleaning and bathing that are polluted by the discharge of stormwater;
(4) body and clothing of those walking through floodwaters, which in turn can lead to further dispersion of pathogens and spread of disease.

There are many pathogenic bacteria that may cause outbreaks of diarrhoea and gastro-enteritis during and after flood events caused by ingestion of bacteria from contaminated water. The most common pathogens in the faecal–oral transmission route are pathogenic *Escherichia coli*, but there are also risks for outbreak of other diseases such as cholera and melioidosis.

2.2.2 Helminth parasitic infections

Faecally contaminated wet soils are ideal conditions for spreading of intestinal nematode worm infections such as roundworm (*Ascaris lumbricoides*), whipworm (*Trichuris trichiura*) and hookworm (*Ancylostoma duodenale or Necator americanus*). Flooding may disperse helminth eggs in the environment and provide conditions that are conducive for the development of these helminths.

The best way to manage parasite problems is to break their life cycle by destroying their breeding habitat. In practical terms this means ensuring that soils do not remain wet due to regular flooding. Drainage interventions can result in dramatic reductions in the incidence of infection in communities infested with helminths. For example, in Salvador, Brazil, Moraes *et al.* (2003) found that a reduction in flooding reduced the incidence of roundworm by a factor of two and hookworm by a factor of three.

2.2.3 Diseases transmitted by mosquitoes

Undrained runoff and floodwaters provide breeding habitats for various species of mosquito, which may act as vectors for the transmission of the diseases listed below. Strategies to control these diseases need to take into account the nature of the breeding sites of different types of mosquito and the specific nature of the particular mosquito species in the region. Dengue, for example, is an important urban problem in Latin America and in South and South-east Asia, but not in Africa. On the other hand, urban malaria is a major problem in much of South Asia and Africa, but is rare in other continents (Lines 2002).

Urban yellow fever

The *Aedes aegypti* mosquito transmits the virus that causes urban yellow fever and the virus is carried either from one animal to another (horizontal transmission) by a biting mosquito (the vector) or the virus is passed to its offspring via infected eggs (vertical transmission). The mosquito eggs are resistant to drying and lie dormant during dry conditions. These hatch when the rainy season begins, which ensures that transmission continues from one year to the next.

Dengue fever and dengue haemorrhagic fever

Dengue (and dengue hemorrhagic) fever is a tropical disease caused by one of four different viruses which have a life cycle that involves humans and *Aedes aegypti*, a domestic, day-biting mosquito that feeds on humans. It is one of the most important mosquito-borne viral diseases affecting humans. Its global distribution is comparable to that of malaria, and an estimated 2.5 billion people live in areas at risk for epidemic transmission.

The primary activity of Dengue control programmes is the control of larval habitats of *Aedes aegypti* mosquitoes in order to reduce the adult mosquito population. The existence of breeding sites is related to specific human behaviours (individual, community and institutional), since any receptacle capable of holding clean rainwater is a potential breeding site for *Aedes* eggs. Therefore, it is important to recognise that strategies to encourage communities to eliminate breeding sites for the *Aedes* mosquitoes, such as water containers, old tyres, urns and other water recipients, are generally more important to control the disease than investments in drainage infrastructure (Lloyd 2003).

Filariasis

Filariasis is a chronic worm infection that causes the disfiguring and disabling disease elephantiasis. It is transmitted by a variety of mosquito species but commonly by *Culex quinquefasciatus* which occurs in almost all African towns. It is well adapted to urban conditions, largely because the polluted water in which it breeds is abundant in urban areas (e.g. pit latrines, soakage pits, septic tanks and blocked drains polluted with domestic refuse) (Lines 2002).

Malaria

Plasmodium protozoa cause malaria, which is considered to be the most important mosquito-borne disease. The protozoon is transmitted to humans by a variety of *Anopheles* mosquitoes. Although generally associated with rural areas, malaria is an increasing problem in peri-urban areas both during and after the rainy season. As mosquitoes that transmit malaria breed in clean water, poor drainage of unpolluted runoff is important for mosquito breeding sites. The fundamental objective of drainage for malaria control is to drain ponded water before larvae can mature within the duration of the mosquito breeding cycle (1 week or less).

2.2.4 Diseases transmitted by rats and snails

Leptospirosis

Leptospirosis (Weil's disease) is a bacterial disease transmitted to humans through contact with water, food, or soil containing urine from infected animals.

Rats are the most common vectors for the disease. Outbreaks of leptospirosis are usually caused by exposure to water contaminated with the urine of infected animals and urban floods may cause epidemics of leptospirosis. Improper housing in flood-prone areas, poor drainage of rainwater and inadequate garbage disposal increases the vulnerability to the disease.

Schistosomiasis

Schistosomiasis is caused by flukes, whose complex life cycles involve specific fresh-water snail species as intermediate hosts. The snails breed in stagnant water and, if infected, release large numbers of minute, free-swimming larvae (cercariae) capable of penetrating the unbroken skin of the human host. Even brief exposure to contaminated water (e.g. by wading or bathing) may result in infection and outbreaks of schistosomiasis may occur when endemic areas become flooded.

2.3 VULNERABILITY AND LIVELIHOOD IMPACTS

Millions of people across the world suffer every year from the misery and danger of flooding in their streets and homes. For some, the flood washes away the savings of a lifetime in only a few hours; for others, long-term ponding during the rainy season is simply accepted as the fate of being poor (Kolsky *et al.* 1992).

The urban poor also suffer from a wide range of environmental problems related to where they live. Low-income communities living in slums next to drainage channels are especially susceptible to environmental health problems including those related to flooding (see Figure 2.4). The health impact of diseases transmitted by various mosquitoes is a major problem, which causes residents of tropical cities, in particular those living in poor housing, to spend hard-earned cash on repellent coils and sprays (Lines 2002).

The risks of flooding are not evenly distributed across cities or across society, and the vulnerability of a community prone to environmental-related hazards is not homogenous. In other words, disaster impacts and social responses are likely

Impacts of flooding on society 25

Figure 2.4 Low-income communities in Cirebon, Indonesia live under precarious living conditions and are prone to the impacts of flooding and poor drainage. (Photo: Martin Strauss, SANDEC.)

Figure 2.5 Urban poor living on the banks of an urban river in Jakarta, Indonesia. (Photo: SANDEC.)

to vary from one community sub-group to another. The differences in vulnerability, are influenced by an understanding and responses to flood warnings which may result in different levels of impacts upon different social groups of society – known as *social amplification* of disasters (Affeltranger 2001).

In addition to other factors, social amplification is related to where poor communities live. Marginal land of less commercial value is often the only land where the urban poor can afford to live. Poorer households are much more likely to inhabit environmentally vulnerable areas, such as those which are at risk from flooding, landslides and pollution. Figure 2.5 shows a typical picture seen in many

Poverty

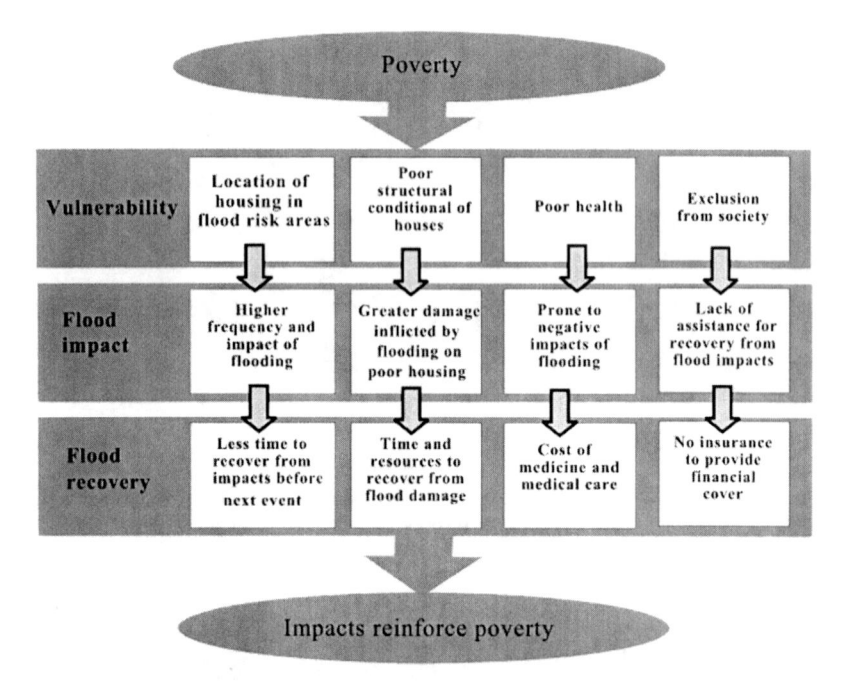

Vulnerability	Location of housing in flood risk areas	Poor structural conditional of houses	Poor health	Exclusion from society
Flood impact	Higher frequency and impact of flooding	Greater damage inflicted by flooding on poor housing	Prone to negative impacts of flooding	Lack of assistance for recovery from flood impacts
Flood recovery	Less time to recover from impacts before next event	Time and resources to recover from flood damage	Cost of medicine and medical care	No insurance to provide financial cover

Impacts reinforce poverty

Figure 2.6 Livelihood impacts of flooding reinforce conditions of poverty.

cities in developing countries where poor families live next to watercourses and are therefore at high risk of flood damage.

Communities who live on steep hillsides are also vulnerable to the problems related to landslides caused by inadequate drainage. Squatter settlements are often constructed on deforested hillsides with cheap rent as a consequence of the urban land ownership. As a result of torrential rains, landslides in these vulnerable areas can cause significant damage and kill many people each year (Nascimento et al. 1999).

The location of poor neighbourhoods and inferior construction materials used to build the homes are other reasons for greater vulnerability. In addition, poor families may not learn about impending disasters or evacuation plans because of illiteracy or the lack of telephones, radios or televisions, and a lack of transportation prevents poor households from moving themselves and their possessions out of harm's way.

Given the vulnerability of the urban poor in relation to where they live, poorer families are often most affected by floods. However, these people have the least amount of resources (otherwise referred to as assets) to recover from the impact. Figure 2.6 illustrates how a combination of a greater vulnerability to flooding and lack of resources to recover from flooding compounds upon the urban poor and may exacerbate conditions of poverty.

Thus, a combination of direct and the indirect impacts have both social and economic implications that compound especially on the urban poor. These groups are more vulnerable as they live in precarious situations that are poorly served by urban infrastructure and services and they suffer the most due to loss of employment, possessions and housing.

Although there are implications of flooding at the societal level, there are also many impacts at the household and local community level. Social groups that are particularly vulnerable to adverse effects of flooding include children and the elderly as well as other disadvantaged groups such as physically disabled people who experience particular difficulties in dealing with disasters. The problems of poor drainage and flooding of domestic properties tend to disproportionately affect women. Women often have to deal with the social and emotional upheaval that comes from dealing with death, disease and food shortages that may occur in the aftermath of floods. In addition to this, cultural practices in some countries dictate that women must always be escorted in public by male relatives, which may increase women's vulnerability in flood events (Francis 2002).

2.4 PERCEPTIONS OF AND RESPONSES TO FLOODING

The relative impacts of flooding compared with other environmental hazards have a direct bearing on the way floods are perceived and how communities respond to them. The example described in Box 2.1 provides an example of a flood-prone urban community in India who developed coping strategies for dealing with the consequences of inundation. Although the consequences of flooding may sometimes be devastating, especially for poor communities who lack the financial resources to rebuild their homes, it illustrates that flooding is often accepted as a fact of life for these vulnerable communities. The benefits of living close by employment opportunities and urban services usually outweigh the disadvantages associated with increased flood risk, which is often perceived as a natural and seasonal event (Stephens *et al.* 1996).

In many situations, these communities have little option about where to live, due to the economic pressures of poverty. However, a study of the environmental hazards and risks to livelihoods of the poor in informal settlements in Indonesia indicated that they worry about the risk of disasters such as large-scale flooding (Santosa 2003).

As a response to these threats and in order to mitigate the risks, many communities develop complex indigenous flood-response strategies to adjust to the flood situation, such as warning systems to give people notice to evacuate their homes and find shelter from floods. These systems may appear to be loosely organised, but in practice they are often effective for disseminating basic warnings over wide areas (Cuny 1992). They are more common in rural communities but often remain in urban environments, if only in part, and may be significant in relation to the response to flood warnings (Schware 1982).

> **Box 2.1 Perceptions of flooding and drainage interventions in Indore, India**
>
> Low-income residents who live in flood-prone areas in Indore in Madhya Pradesh, India perceive flooding to be a natural event and this helps to understand why they tolerate the negative impacts and accept it as a part of life.
>
> In urban slums, flooding is generally ranked low in comparison to other risks and problems, such as improvements in job opportunities, provision of housing, mosquitoes and smelly back lanes. In some areas, residents felt that the after effects from flooding and inundation, such as stale water, contaminated mud and noxious odours, were more important than the immediate effects, which may include water entering homes and, in riverbank areas, loss of possessions. The risks of flooding during the monsoon season are borne as part of a trade-off of the risks and benefits, social and economic, of living in a flood-prone area. These benefits include centrality in the city, access to sources of employment, low land costs, access to services, well-developed social support networks and safety for children. As a result, although flooding is a stress, and may have problematic effects, its relative significance is reduced by a general acceptance of flooding as a natural phenomenon necessary for agriculture.
>
> A 'hazards culture' has therefore developed in flood-prone areas, with coping mechanisms seen as part of usual activities. Residents have developed rudimentary but effective flood prediction (based on the duration and intensity of rain) and protection systems, contingency plans for evacuating persons and possessions, and ways of mobilising support and assistance in local communities. During the monsoon season, families are prepared to move the elderly, children and animals to higher ground in response to the threat of severe floods. Many households own large cases and other forms of luggage ready in case they need to quickly pack their most valuable possessions in order to carry them away.
>
> Engineering interventions are perceived to disturb natural flood patterns and maintenance is also seen to influence the ability of these communities to predict flood events, which may disrupt communities' ability to respond to the imminent threat of a flood. As a consequence, these interventions are not always well tolerated and it is important for drainage engineers to take these into consideration. For example, the predictability of changes in water levels and the rate of rise in water levels are/were considered by residents to be more important than the actual duration or depth of flood events.
>
> *Source*: Stephens *et al.* 1996.

However, in other situations urban communities may lack this innate knowledge, either due to the fact that they are immigrants to the area or because the environment has changed so much that their ability to predict a flood situation is lost. In the example described in Box 2.2, the President of Brazil, Luiz Inácio Lula da Silva, recounts his childhood experiences of life in a *favela* (slum) in

> **Box 2.2 Responses to flooding in São Paulo, Brazil**
>
> "When our house flooded, I sometimes woke up at midnight to find my feet in water, cockroaches and rats fighting over space, and various objects floating around the living room. Our biggest concern was preventing the furniture from getting ruined. Not that we had much to get ruined as we didn't even have a television or a fridge. But we had to lift up the bed and the wardrobe, and get our 60-year-old mother out, carrying her on our shoulders so our dear old lady didn't get soaked. And then we had fun. It was actually great because everyone went to a club called Ponte Preta, where we turned the disaster into a social event, the young people danced, had fun and played table tennis, while the older people commiserated over their misfortune. But we still found something useful to do. A group of us used to get an inner-tube from a lorry tire, blow it up and use it like a boat – we floated around visiting all the houses to see whether there was anything we could do, like lifting up the fridge or cooker, or helping to get elderly people out. So, when we were younger, we managed to transform the suffering into a fun event. But I know that other people suffered."
>
> *Source*: da Silva *et al.* 2003.

São Paulo in relation to how the community responded to the flood situation. Of specific interest in this example is the fact that the flood event was viewed differently by different members of the community and the fact that they worked together communally to help those more disadvantaged, such as the old and disabled.

2.5 IMPLICATIONS ON THE DESIGN OF URBAN STORMWATER MANAGEMENT SYSTEMS

An appreciation of the fact that many residents often have an ambivalent attitude towards flooding due to the trade-offs between living in a location that provides access to services and other opportunities to support their livelihoods helps to explain the logic of 'living with floods' rather than attempting to prevent them through the provision of large-scale engineering interventions (Few 2003). There are also important considerations related to the level of poverty, which affect where people live, thus making the poor more vulnerable to flooding and their resources to be able to cope with and recover from a flood event. These are important considerations to be taken into account in the design of urban drainage systems and flood mitigation strategies.

Thus local residents' perceptions of flooding and their responses to flooding should influence the design of the most appropriate form of drainage intervention and the development of flood mitigation and response strategy. For instance, the example described above in Box 2.1 suggests that engineered interventions aimed at reducing flooding need to take into account local perceptions of flooding as well as the coping strategies developed by individuals and communities.

A livelihoods approach (Ashley and Carney 1999; Meikle *et al.* 1999) is one way to approach the complexities of the social environment within cities that can help understand the full complexity of the flood impacts and why they are perceived in the way they are. In the case of flooding, there are short-term livelihood impacts such as structural damage to the homes, resulting in a direct impact on the asset base of the poor. There are also many indirect effects such as the loss of working days required to repair structural damage or the increased prevalence of illness causing poor families to redirect limited financial resources towards medical treatment.

Livelihood strategies are combinations of activities in which the poor utilise their resources to ensure their survival in the face of external factors such as those environmental changes caused by flooding. The organisational capacity of communities during flood events is important as it increases the chances of greater self-reliance and preparedness amongst households and neighbourhoods to cope with flooding. From a livelihoods perspective, these life-saving actions are only likely to come about after years of community activities, which address a broader range of development issues concerning urban communities (Sanderson 2000).

It is important to recognise that many health problems related to diseases associated with poor drainage can be avoided without investments in large-scale drainage infrastructure. In many situations, investments in smaller-scale infrastructure for drainage at the household and community level, combined with health education so that people understand the health risks of poor drainage, may be a more cost-effective and effective means of improving problems associated with poor drainage than investments in large-scale flood alleviation infrastructure.

It is therefore important not to assume that drain construction will always be the most appropriate and cost-effective form of intervention to resolve urban drainage problems. These are often necessary, but in combination with other interventions designed to improve environmental health. For instance, it is important that engineers understand the importance of mosquito control in construction sites, which often serve as major breeding areas. The removal of ponds and small puddles is important both during and after construction work as an effective of mosquito control measure (Kolsky 1999).

Also, drainage engineers require greater insight into the socio-economic implications of their designs based on technical parameters and should place greater consideration of the importance of stakeholders' involvement in decision-making and planning. Thus, the importance of non-structural strategies resulting in changes to behaviour and responses to environmental threats may result in greater benefits than those associated with the construction of drains. The participatory approaches described later in the book are ways in which the livelihood issues may be incorporated in the planning and design of urban drainage and flood mitigation responses.

2.6 REFERENCES

Affeltranger, B. (2001) *Public Participation in the Design of Local Strategies for Flood Mitigation and Control.* Technical Documents in Hydrology No. 48. International Hydrological Programme, UNESCO, Paris.

Andjelkovic, I. (2001) *Guidelines on Non-structural Measures in Urban Flood Management.* IHP-V Technical Documents in Hydrology No. 50. International Hydrological Programme, UNESCO, Paris.

Ashley, C. and Carney, D. (1999) *Sustainable Livelihoods: Lessons from Early Experience.* Department for International Development, London.

Cairncross, S. and Ouano, E. A. R. (1991) *Surface Water Drainage for Low-income Communities.* WHO/UNEP, World Health Organization, Geneva, Switzerland.

Cuny, F. C. (1992) Living with floods: alternatives for riverine flood mitigation. In: *Managing Natural Disasters and the Environment* (eds. A. Kreimer and M. Munasinghe), World Bank, Washington DC, 62–73.

da Silva, L. I. L., de Castro, C. R., de Fátima Machado, S., de Orato Santos, A. O., Ferreira, L. T. T., Teixeira, P., Suplicy, M. and Dutra, O. (2003) The programme for land tenure legalization on public land in São Paulo, Brazil. *Environment and Urbanization* 15(2), 191–200.

Few, R. (2003) Flooding, vulnerability and coping strategies: local responses to a global threat. *Progress in Development Studies* 3(1), 43–58.

Francis, J. (2002) *Understanding Gender and Floods in the Context of IWRM.* Gender and Water Alliance. November 2002. http://www.genderandwateralliance.org/reports/discussion_paper_on_gender_and_floods_by_JF.doc Accessed 2nd October 2004.

Jahan, S. (1988) Coping with flood: the experience of the people of Dhaka during the 1988 flood disaster. *Australian Journal of Emergency Management* 15(3), 16–20.

Kolsky, P. (1999) Engineers and urban malaria: part of the solution, or part of the problem. *Environment and Urbanization* 11(1), 159–163.

Kolsky, P., Hirano, A. P. and Bjerre, J. (1992) *Directions in Drainage for the 1990s.* UNDP World Bank Regional Water and Sanitation Program & World Health Organization, South East Asian Regional Office, New Delhi.

Lines, J. (2002) How not to grow mosquitoes in African towns. *Waterlines* 20(4), 16–18.

Lloyd, L. S. (2003). *Best Practices for Dengue Prevention and Control in the Americas.* Strategic Report 7. Environmental Health Project, U.S. Agency for International Development, Washington, DC.

Meikle, S., Ramasut, T. and Walker, J. (1999) *Sustainable Urban Livelihoods: Concepts and Implications for Policy.* Development Planning Unit, University College London.

Moraes, L. R. S., Azevedo Cancio, J., Cairncross, S. and Huttly, S. (2003) Impact of drainage and sewerage on diarrhoea in poor urban areas in Salvador, Brazil. *Transactions of the Royal Society of Tropical Medicine and Hygiene* 97(2), 153–158.

Nascimento, N. O., Ellis, J. B., Baptista, M. B. and Deutsch, J.-C. (1999) Using detention basins: operational experience and lessons. *Urban Water* 1(2), 113–124.

Sanderson, D. (2000) Cities, disasters and livelihoods. *Environment and Urbanization* 12(2), 93–102.

Santosa, H. (2003) *Environmental Hazards Management in Informal Settlements to Achieve Sustainable Livelihoods of the Poor: The Case of East Java, Indonesia.* http://www.csir.co.za/akani/2003/mar/environhazard.pdf CSIR Building & Construction Technology, Pretoria, South Africa. Accessed 2nd October 2004.

Schware, R. (1982) Official and folk flood warning systems: an assessment. *Journal of Environmental Management* 6(3), 209–216.

Stephens, C., Patnaik, R. and Lewin, S. (1996) T*his is my beautiful home: risk perceptions towards flooding and environment in low-income urban communities: case study in Indore, India.* London School of Hygiene and Tropical Medicine, London.

Tapsell, S. (1999) *The Health Effects of Floods – The Easter 1998 Floods in England.* Article Series 3/99. Flood Hazard Research Centre, Middlesex University, UK.

UNCHS (1993) *Maintenance of Infrastructure and its Financing and Cost Recovery.* United Nations Centre for Human Settlements, Nairobi, Kenya.

WHO (2002a) *Floods: Climate Change and Adaptation Strategies for Human Health.* Report on a WHO meeting London, United Kingdom 30 June – 2 July 2002, WHO Report EUR/02/5036813. WHO Regional office for Europe, Denmark.

WHO (2002b) *Flooding: Health Effects and Preventive Measures.* WHO Fact Sheet 05/02. Copenhagen and Rome, 13th September 2002.

3
Integrated framework for stormwater management

Integrated urban water management (IUWM) incorporates principles of integrated water resource management (IWRM) in order to develop solutions for the specific challenges related to the management of water systems within the context of the urban environment. This chapter describes how these principles may be applied to runoff control and flood mitigation strategies, and three important components of an integrated framework for urban drainage are described. These relate to the management of the urban environment and the land–water interface, co-ordination of infrastructure and service provision, stakeholder participation and institutional partnerships.

3.1 THE NEED FOR AN INTEGRATED FRAMEWORK

The widespread degradation of water resources caused by pollution from urban runoff combined with the impacts of flooding and poor drainage of runoff has led to a critique of traditional approaches for the design and operation of urban drainage systems. One of the main criticisms of conventional approaches for the management of stormwater is that they focus on technical solutions in a fragmented manner and

© 2005 IWA Publishing. *Urban Stormwater Management in Developing Countries* by Jonathan Parkinson and Ole Mark. ISBN: 1843390574. Published by IWA Publishing, London, UK.

that they tend to address problems only as they occur with little attempt to prevent or mitigate problems in advance.

Many of the problems described in Chapters 1 and 2 support the argument for the need to adopt an integrated approach in urban stormwater management practices and these are summarised below:

(1) *Lack of consideration of complexities of environmental management*: Agencies responsible for stormwater management are increasingly being called upon to address water quality and natural resources issues in addition to their traditional focus on drainage of runoff and flood protection.
(2) *Lack of institutional co-ordination and management operations*: Lack of co-ordination and integration between urban stormwater management activities and other urban services can affect the sustainable and cost-efficient operation of urban drainage systems. For example, solid waste frequently ends up in the stormwater drains, but refuse collection services are generally not co-ordinated with drain-cleaning activities.
(3) *Insufficient consideration of the interactions between communities and drainage systems*: Drainage channels often provide a range of additional uses for local communities. For example, they may be used as sources of water for irrigation or they may be utilised as a source of low-grade water for various domestic purposes and recreational activities (see Figure 3.1).
(4) *Conventional centralised decision-making processes tend to be unresponsive to the concerns of local stakeholders*: In order to take the demands from these stakeholders into account, a more inclusive and participatory approach to urban water planning is required.

As a result of the above considerations, there is increasing recognition that stormwater runoff needs to be managed within an integrated framework of water resource management, which emphasises the need for integrated planning, and the interactions between land and the aquatic environment.

Figures 3.1 Local residents in Vientiane, Lao PDR, use drainage channels for (a) washing vegetables and (b) fishing. (Photos: Birgitte Helwigh.)

3.2 KEY ELEMENTS OF INTEGRATED WATER RESOURCE MANAGEMENT (IWRM)

IWRM aims to overcome the many water-related problems that arise in society as a result of the uncoordinated and inefficient use, and abuse of water resources. IWRM focuses on the development of coherent and comprehensive policies, and legal instruments for co-ordination and regulation of activities that affect the availability, distribution and quality of water resources. The Global Water Partnership (GWP) summarises IWRM as "a process that promotes the co-ordinated development and management of water, land and related resources to maximise resultant economic and social welfare in an equitable manner without compromising the sustainability of vital ecosystems" (GWP 2000).

The potential benefits of adopting IWRM approaches may be measured in terms of both direct and indirect improvements to environmental quality – particularly for those who live in areas affected by pollution and have the most to gain from water quality improvements. IWRM has adopted a number of principles based on the Dublin Statement on Water and Sustainable Development (see Box 3.1), which was the main outcome from the *International Conference on Water and the Environment* held in Dublin in 1992 ICWE (1992). Taking the Dublin principles into consideration, the conceptual framework and operational structure of IWRM promotes the strategic objective to balance *water for livelihoods* and *water as a resource* (see Figure 3.2).

The IWRM conceptual framework may be applied within the context of an urban catchment and provides an approach towards planning and design of stormwater management strategies, with a specific consideration of issues related to:

- human health,
- environmental protection,
- natural resource management.

Box 3.1 The principles of IWRM

Environmental sustainability: this relates to the principle that fresh water is a finite and vulnerable resource, essential to sustain the environment and human livelihoods.

Economic efficiency: this principle recognises that water has an economic value and should be therefore be managed as an economic good. This also encourages the need for strategies for water conservation and reuse.

Social equity: There are two principles in the Dublin Statement that relate to the equitable distribution of water across different social and economic groups. Firstly, water management should be based on participatory approaches which involve stakeholders at all levels, and secondly, that women play a central role in the provision, management and safeguarding of water resources.

[1] Based upon the Dublin Statement on Water and Sustainable Development.

Figure 3.2

Principles: Economic efficiency — Social equity — Environmental sustainability

Structure:
- Enabling environment: Policies, Legislation, Economic instruments
- Institutional framework: Agencies for water resources management, Public-private partnerships, Stakeholder participation
- Management instruments: Monitoring, Evaluation and assessment, Regulation

Objective: Balance of *water for livelihoods* and *water as a resource*

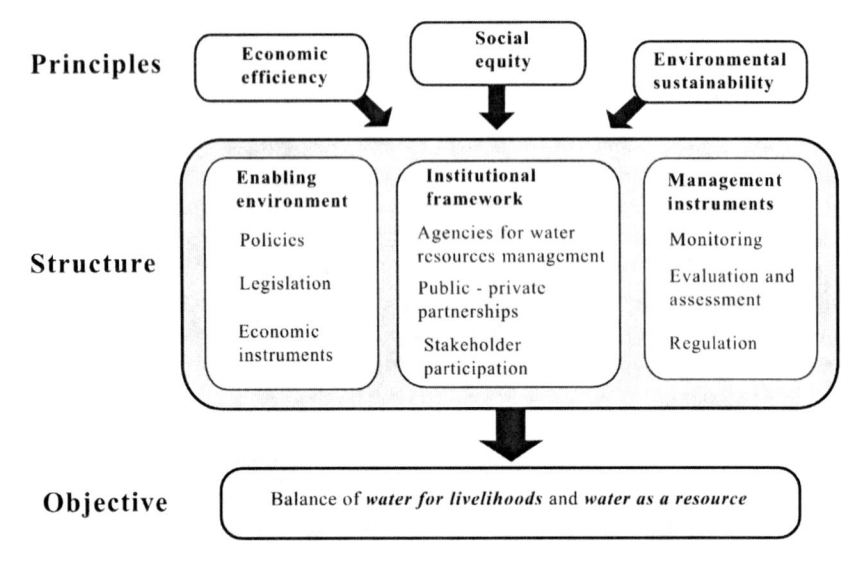

Figure 3.2 Conceptual framework and operational structure for IWRM.

The IWRM framework illustrated in Figure 3.2 recognises that complementary elements of an effective water resources management strategy must be developed and strengthened concurrently. This may be achieved via the following instruments:

(1) *Enabling environment*: the framework of national policies, legislation and regulations for water resources management.
(2) *Institutional framework*: the function, roles and responsibilities of various agencies and organisations at different levels of administration.
(3) *Management instruments*: the operational instruments and administrational mechanisms for effective implementation, monitoring and regulation of policies.

3.2.1 Integrated urban stormwater management

Table 3.1 summarises the short-, medium- and long-term objectives of stormwater management strategies. In the short term, the priorities are runoff control flood protection and pollution mitigation strategies which in many developing countries, has yet to be addressed effectively. The medium-term objectives focus on the development and implementation of water quality improvements, water conservation and strategies to preserve the hydrology of the natural catchment.

The longer-term objectives place greater emphasis on preservation of natural resources and the amenity value of water in the urban environment for recreational activities, and to promote an increased awareness of environmental issues. These additional objectives involve a broadening of traditional approaches towards the

Table 3.1 Objectives of stormwater management strategies.

Short-term	
Flood protection	Reduce incidence of flooding, prevent structural damage to properties and infrastructure, mitigate flood hazards, and minimise disruption and other risks to human life.
Environmental health protection	Mitigate deteriorations in environmental health – eliminate standing water and the consequences of danger from mosquito-related diseases.
Erosion and sediment control	Control erosion and reduce instability of hillsides, reduce soil loss from construction areas in order to reduce downstream sediment-related problems.
Medium-term	
Pollution prevention, control and mitigation	Preserve and/or enhance the receiving water ecosystem. Protect water quality via the reduction of discharges of pollutants into the environment.
Water conservation	Regeneration of natural surfaces waters (rivers and lakes) and groundwater recharge; promote reuse through rainwater harvesting.
Preservation of natural hydrology	Reduce modifications to natural streams, preserve natural drainage pathways and floodplains; replicate natural flow regimes so that storm runoff hydrographs resemble pre-development patterns.
Long-term	
Amenity	Integrate urban drainage into landscape design and adopt sustainable principles to create a healthier environment, promote land- and water-based recreational activities, and stakeholder satisfaction.
Protection of natural habitats	Protection of biodiversity through the preservation and restoration of natural habitats for flora and fauna.
Resource conservation	Minimise energy consumption and other natural resources associated with the construction and operation of drainage infrastructure.

planning and design of urban drainage systems involving the development of multi-purpose drainage systems that incorporate existing natural drainage channels, streams, and rivers with features of habitat protection and recreation (see Chapter 7).

Although these may initially appear to be somewhat idealistic goals, especially considering the existing situation in many developing countries, it is important that planners and designers of urban drainage systems aim to satisfy the needs of future generations within keeping with the objectives of sustainable development as defined by the World Commission on Environment and Development in 1987.

An example of how these objectives may be put into practice has been initiated in Malaysia. Box 3.2 describes an innovation in stormwater and drainage practices

> **Box 3.2 Development of the stormwater management master plan in Ipoh, Malaysia**
>
> In the city of Ipoh, Malaysia, due to the unsuccessful efforts in the past to solve the water quantity and quality issues in urban areas, the authorities responsible for urban stormwater management realised that the only way to approach the problems of control of runoff and pollution was to adopt an integrated approach towards urban development. The criteria used for selecting an option included total cost, social and environmental impact, constraints against implementation, etc. Only by considering the full cost attributable to the project, including external costs of downstream works and environmental protection costs, it is possible to derive the most suitable long-term option. A stormwater management plan was developed for Ipoh as a framework for action in order to provide direction for stormwater management practices. The plan provides a framework for achieving goals by identifying sustainable stormwater programmes and solutions, which can be implemented as a co-operative approach by stormwater managers and the community. The primary outcome of the stormwater management planning is identification of management actions, which are tailored to the specific stormwater issues identified. The plan will therefore provide a valuable management tool for the Council, stormwater managers and community, which will help to protect the environmental, social and economic values of Ipoh city.
>
> *Source*: Hashim and Al-Mamun 2002

in the city of Ipoh where a stormwater management master plan has been developed on the basis of principles of IUWM. More details about recent developments in Malaysia relevant to the development of sustainable urban drainage systems are found in Chapter 7.

In response to the problems presented at the beginning of the chapter, three fundamental components can be identified that form the basis of a stormwater management strategy which meets objectives of IUWM.

As described below and as illustrated in Figure 3.3, these are:

(1) *Catchment planning*: land control strategies, water resource management and consideration of the land–water interface.
(2) *Urban infrastructure and services*: co-ordination of services for drainage, solid waste management and sanitation, and co-operation between agencies responsible for different services.
(3) *Stakeholder participation*: involvement of local stakeholders in decision-making and planning processes, and partnerships in the implementation and management of urban services.

Referring to the integrated approach towards urban water management described above, the following sections describe in more detail how these aspects apply to the management of urban stormwater.

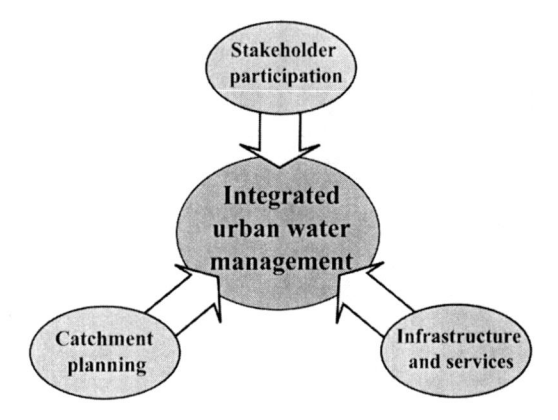

Figure 3.3 Main components of IUWM.

3.3 INTEGRATED PLANNING OF URBAN CATCHMENTS

An IUWM approach for stormwater management requires greater emphasis on urban drainage as an integral component of urban development and recognises the importance to ensure that urban water systems do not impose undue harm on local water resources or downstream watercourses. The recognition of the interactions between stormwater runoff and land-based activities – both in terms of quantity and quality – requires a holistic approach in which aspects of land planning and site design consider the impact of development on the hydrological cycle. This requires a need to control urban developments and ensure that natural drainage paths and their functions are included in urban development plans.

Box 3.3 describes a research project in Vientiane, Lao PDR, which set out to develop strategies based on IWRM principles to reduce the impacts of drainage on the natural hydrological cycle and improve the quality of the receiving waters.

Box 3.3 Integrated urban drainage system design in Vientiane, Lao P.D.R

Vientiane's population of 0.5 million is relatively small, especially compared to other capital cities in Asia, but it suffers from many problems that are common throughout urban areas in South-East Asia. Traditionally, residents lived adjacent to an intricate network of densely vegetated earth-channels and wetlands, and for many these have been an important part of their livelihood strategies – particularly for fish and vegetable production. Even today, the large That Luang Marsh stretching 20 km^2 along the eastern end of the city provides a significant portion of the capital's food produce, as well as revenue for those who work there.

Figure 3.4 Hong Ke channel, part of the natural drainage network of Vientiane (a) in 1990 and (b) in 2003 after rectification (Photo: Jean Lacoursière and Lena Vought).

However, although these traditional practices remain, natural drainage channels are deepened and often lined with concrete as part of a strategy to remove stormwater from urban areas as quickly as possible (see Figure 3.4). Nong Chan, the most central wetland of the city, was literally divided in two by one of these channels, inevitably resulting in further loss of natural attenuation and purifying capacity of the drainage network. As budgetary restrictions left sanitation improvements to a later phase of the city infrastructure development programme, the newly rectified network operated under extreme conditions – low flows dominated by sewage, punctuated by stormwater surges during the rainy season – leading to an abrupt increase in nutrients leaving the city.

In order to gain an improved understanding of the existing situation and develop integrated strategies for urban drainage system design, a collaborative research project based on an adaptive environmental assessment and management approach was initiated by Swedish and Laotian scientists in 1999. The work was done in partnership with the Vientiane Urban Development Administration Authority (VUDAA) and the Lao National Mekong Committee (LNMC). This was particularly important as the possibility of low-quality stormwater reaching the Mekong River could raise cross-boundary issues.

Central to the approach was the development of a computer model assessing the effects of selected management options to control flooding and protect receiving water quality. These included replanting drainage channels with vegetation to improve the pollutant retention and rejuvenate purifying capacity as well as greater attenuation of peak flows.

One of the important side benefits of the project was the heightened awareness and willingness of practitioners and local government officials to consider ecologically sensitive design features for urban stormwater management. Although improvements in water quality were considered to lead to a reduction of mosquitoes breeding sites due to the re-colonisation by fish and aquatic invertebrates, a more thorough assessment of the proposed management strategies with regards to health issues was considered to be essential.

Source: Amena *et al.* 2004, Larsson *et al.* 2002

Integrated framework for stormwater management 41

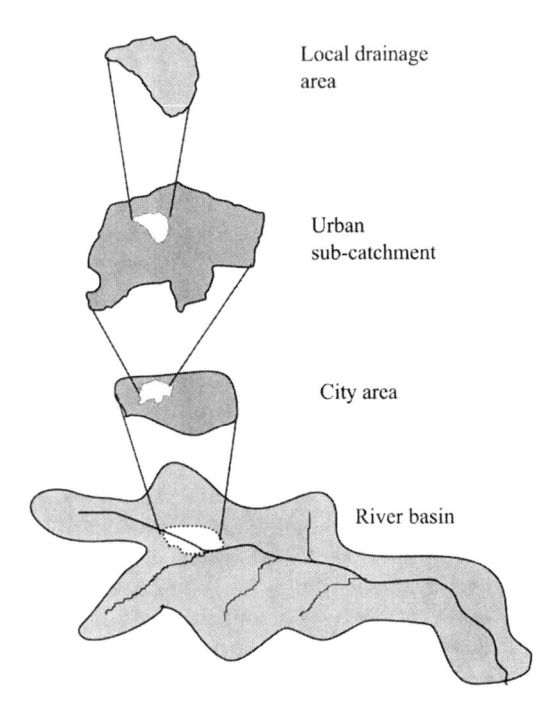

Figure 3.5 Relationships between areas of development, urban catchments and river basin. (Adapted from CWP 1998.)

Although IUWM strategies focus on activities within the urban environment, they should consider the interactions between city-level activities and the influence that these exert on natural water resources in the surrounding hinterland. Both IWRM and IUWM consider the catchment area as the spatial element for the management of water resources and, as illustrated in Figure 3.5, there are four levels of catchment management:

(1) *Local drainage area*: a localised area of urban development, which is served by a 'tertiary' drainage system consisting of small pipes or open channels.
(2) *Urban sub-catchment*: an urban sub-catchment (or district) which is served by a secondary drainage system, which connects to the larger city-level drainage system.
(3) *City area*: served by the city-level (primary) drainage system which consists mainly of large, open channels and urban streams.
(4) *River basin*: the drainage system of the river basin as a whole consists of natural drainage channels, streams and rivers.

Due to these interrelationships, drainage problems in one area are closely interrelated to the responses to drainage problems elsewhere. Better drainage in one neighbourhood means that surface water flows away faster imposing a greater burden on the capacity of the downstream system. At the same time, drainage improvements at the local level may be of little value if water backs up because of the lack of capacity in the downstream system. Therefore interventions to improve urban drainage are best planned in the context of the city as a whole.

A key element related to integrated planning in urban catchments and the control of urban runoff is the emphasis on control of developments in a way in which flooding problems are mitigated. One aspect of this is to enforce development regulations to minimise runoff from new developments by minimising catchment imperviousness or introducing other source control strategies (see Chapter 7). In addition to this is the need to prevent developments in areas that are identified to be particularly prone to flooding or landslides caused by high runoff. These factors are particularly important for cities in developing countries, which are often characterised by a large proportion of informal settlements.

This introduces the concept of *prevention planning* in order to preserve open spaces, especially those areas demarcated as flood zones, where only certain types of land use are permitted. As described in Chapter 10, zoning is a planning tool that urban authorities can use to control the type of development or redevelopment allowed within the city boundaries. However, the successful implementation of prevention planning strategies requires close co-operation between planners and drainage designers from the early stage of land development, and zoning will only work if there is a strong regulatory and enforcement agency.

Another important point mentioned in Chapter 1 related to integrated planning of urban catchments, is the fact that administrational boundaries of local authorities generally do not follow the boundaries of the hydrological catchment. In the example illustrated in Figure 3.6, the urban area managed by the local authority covers three main hydrological catchments (A–C), but the total area of these

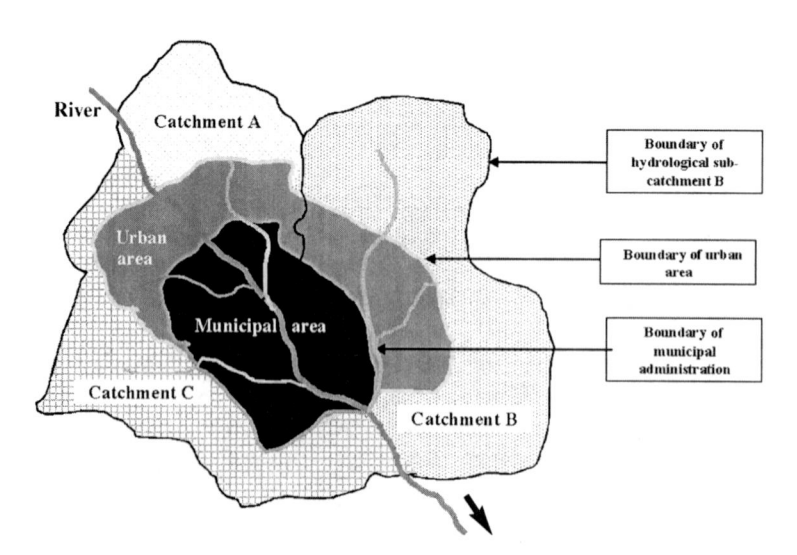

Figure 3.6 Misalignment of catchment hydrological boundaries with the urban area and the area under the control of the municipal administration.

catchments is much larger than the municipal area. Urban developments in areas outside the municipal boundaries contribute towards the urban drainage problems in the city and, due to the linkages between different catchment areas, drainage systems need to be planned and developed in co-ordination with other neighbourhoods and areas of the city.

3.4 INTEGRATED MANAGEMENT OF URBAN INFRASTRUCTURE AND SERVICES

By definition, urban water management involves the management of all forms of infrastructure and services related to water in the urban environment. Therefore, in addition to urban drainage and flood protection, it includes water supply, wastewater and solid waste management systems as well as road systems as these will often be the main drainage channels during large flood flows. IUWM requires taking into account how these components interact and covers both physical infrastructure and management arrangements.

One of the most important opportunities relates to the scope to reduce solid waste in the drainage system, which is particularly important as stormwater drainage systems are frequently blocked by refuse. Integrated management and co-ordination between the relevant agencies also provide greater opportunities for the control of pollution from illegal sewer connections. As shown in Figure 3.7, a combination of these form part of an integrated strategy for improved management of the urban environment.

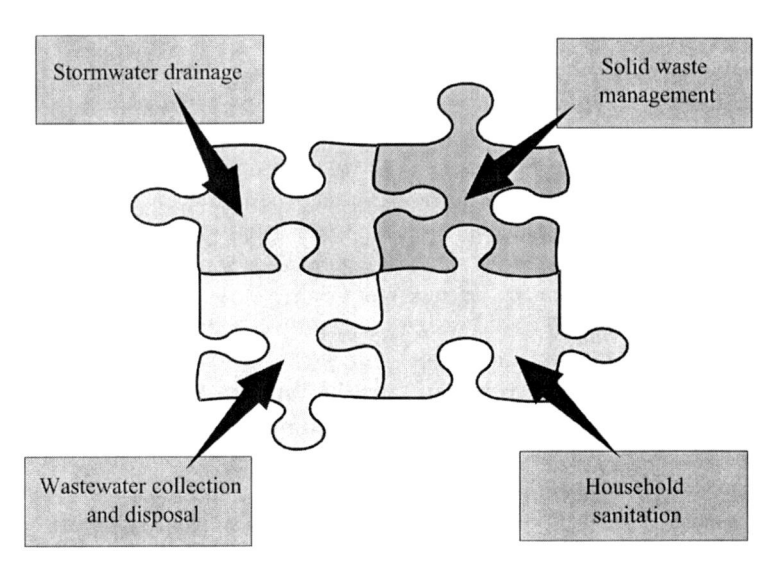

Figure 3.7 Stormwater drainage – an integral component of urban environmental management.

Drainage maintenance departments often lack the equipment to transport solid waste material once it has been removed from the drain. Not only will the solid waste department have the most appropriate equipment for collection and transportation, they will also be able to co-ordinate and plan the collection of drain cleanings so that it does not remain on the roadside after it has been removed from the drain. This is not only a more efficient use of resources, but it also means that the waste is less likely to be washed back into the drain at the onset of the next storm event. This is also particularly important from an environmental health perspective as drain cleanings frequently contain pathogens and are a location for flies to breed.

In addition, it is important that roads are not constructed in a way that fills in and builds upon existing drainage channels, hinder overland flow pathways or constructed on elevated embankments, which cause floodwaters to be impounded as if they were behind a dam. Roads are generally constructed with little consideration for the communities who are at risk from flooding during the rainy season. Therefore, if roads are to function as drainage channels during times of peak flooding (see Chapter 6), the roads department needs to work closely with the drainage department.

The following example describes an innovate project from India which adopted the concept of "roads-as-drains". The project illustrates an integrated approach towards the planning and design of different urban services and infrastructure, with a particular focus on spatial and participatory planning aspects, and an holistic approach towards urban environmental management.

Case study: 'Slum networking' – an integrated approach towards urban upgrading

As noted in Chapter 2, in many cities in developing countries, informal squatter settlements form in low-lying areas and cluster on the banks of drainage channels and natural watercourses. Due to endemic poverty of the local inhabitants and the lack of infrastructure and services, the environmental health conditions are often extremely poor and these settlements frequently become slums. Therefore, there is a close correlation between the slum locations and the natural drainage paths of a city. Studies of several cities in India and in other parts of the world showed that slums are consistently located along these natural paths (see Figure 3.8) (Parikh 2001). These slums are generally considered to be an undesirable element in urban environment, and traditionally urban authorities have tried hard to ignore them – perhaps in the hope that the squalid living conditions would become unbearable for the residents and hence that they would move away.

In India, the *slum-networking* approach has been developed as an innovative approach towards upgrading of low-income settlements, which exploits the linkage between the slums and the natural drainage paths that influence the urban infrastructure and environment of the city (Diacon 1997, Parikh 2001). The approach aims to solve problems of flooding in the city as a whole, whilst concurrently providing services for the slum dwellers in low-lying areas adjacent to natural drainage paths (see Figure 3.9). Consequently, the outcomes are mutually beneficial for both the slum dwellers themselves and the other city residents.

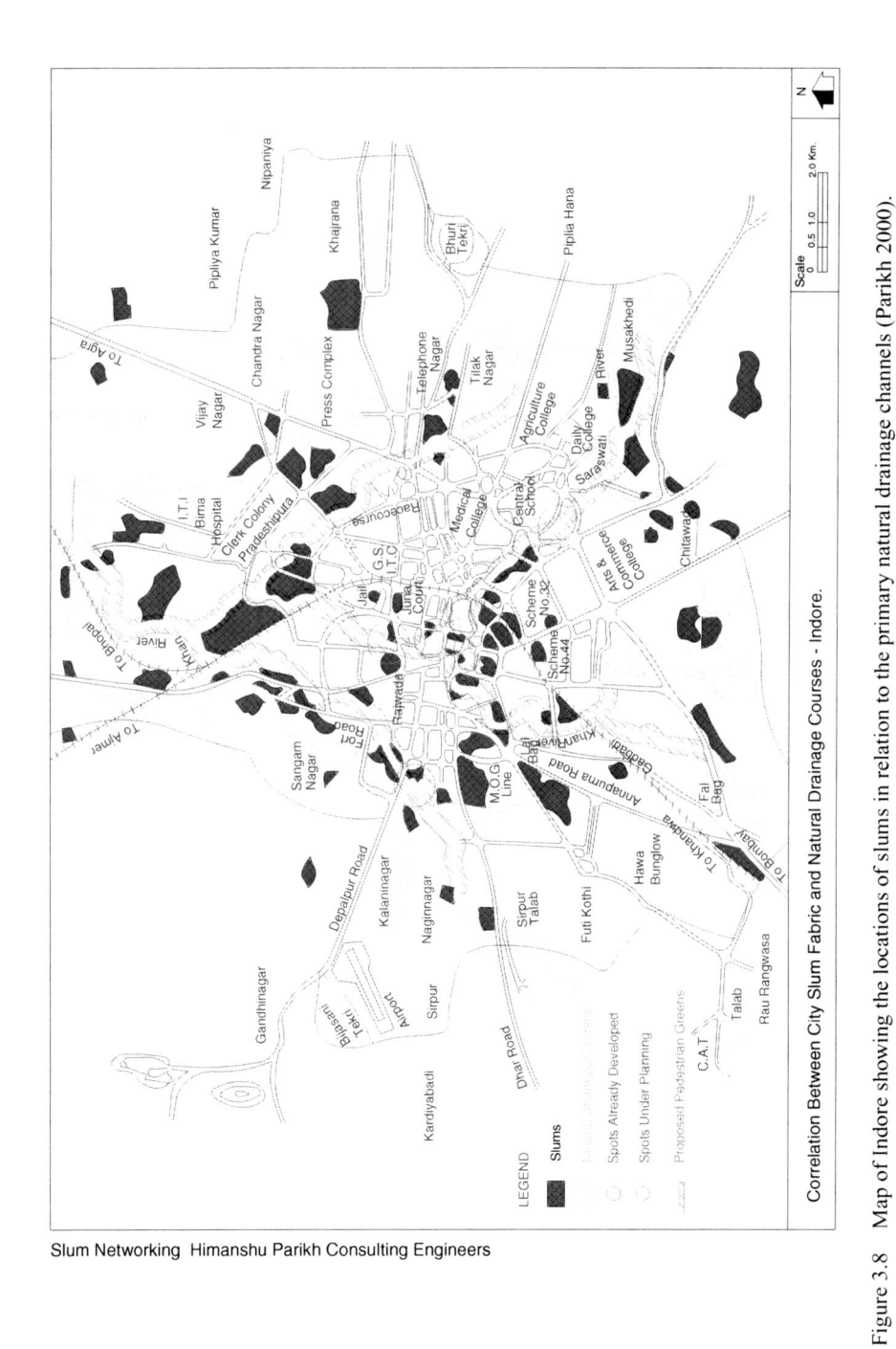

Figure 3.8 Map of Indore showing the locations of slums in relation to the primary natural drainage channels (Parikh 2000).

Figure 3.9 Typical slums in Indore (a) before and (b) after urban upgrading as part of the implementation of the slum-networking project. (Photos: Himachal Parikh.)

One of the advantages of this approach is the fact that the natural drainage paths are the most obvious places to construct drainage channels. The storm drains and sewers can be laid to natural gradients in each settlement resulting in reduced construction costs and improved operational performance. In the slum-networking approach, problems of land acquisition and demolition normally encountered in built-up areas during installation are avoided because the local slum residents are actively involved in the planning process and have a direct interest in the implementation of the project.

Another benefit relates to the cost-effectiveness of the solution. The integrated approach adopted by the project design means that relatively small-scale investments in infrastructure can result in significant benefits. The service infrastructure is simplified and modified so that individual services (instead of shared facilities) can be offered to slum families at low costs. Thus, many solutions that were initially considered to be unviable at micro-level become more economically viable when incorporated within a comprehensive urban development programme. For instance, in Indore, just by providing the missing links between the slums, it was possible to build up city-level sewerage at costs less than half those of the conventional system proposed by Public Health Engineering department (Parikh 2001).

Also related to the cost gains, is the fact that in a networked sewerage system, families other than in slums can also be connected to the same system without recurring the off-site costs (i.e. the cost per family decreases as the contributing families increase). In addition, there are other economies of scale for the project as a whole due to the co-ordination of drainage and road infrastructure. Another way in which the project was able to reduce costs related to the innovative aspects of the engineering design in which the roads were designed so that they would also serve as drainage conduits during flood conditions (the concept of the road as drains is discussed further in Chapter 6).

As communities have the best knowledge and sensitivity of their surrounding environment, the successful implementation of the slum-networking initiative is driven primarily through community participation, in which slum dwellers play an active role in the execution process. At the same time the maintenance burden is reduced and can be shifted from the local government to the individual householders.

As the practice of slum networking evolves from city to city, an increasingly iterative design process is adopted in consultation with the community and, in this process, the role of the external funding agency has shifted from that of a 'benefactor' to a 'catalyst'. Non-governmental organisations (NGOs) play an active role in motivating the communities, mobilising resources from the slum dwellers and converging the efforts of the people with the inputs from the local government and the private sector.

The 'soft underbelly' of the slum-networking project in Indore was that it was a grant and the replicability of the project can therefore be questioned. Furthermore, bilateral grants are normally channelled through government structures and therefore, in spite of the project concept, the project was delivered by the agency and not executed by the community (Parikh 2001). In addition, experience from Indore indicated that the implementation may be problematic and beset by similar problems as more traditional approaches: ineffective solid waste management systems to keep the drains clear from refuse, a lack of demand and willingness to invest by the slum dwellers themselves, as well as problems related to inefficiencies of expenditure and corruption due to the large sums of money involved (Verma, undated).

However, based on the lessons learnt in Indore, the concept was evolved and replicated in demonstration projects in the cities of Baroda, Ahmedabad and Mumbai, each time bringing greater community interaction and self-sufficiency of resources. The transitions from Indore to Baroda and then to Ahmedabad have taken comparatively shorter times in spite of the fact that at each stage the level of self-sufficiency and the degree of community control has increased. These experiences demonstrate that proactive participation of slums dwellers in an integrated programme of urban development can results in significant benefits related to drainage of wastewaters and urban runoff.

3.5 STAKEHOLDER PARTICIPATION AND PARTNERSHIPS

As highlighted in the example from India described above, managing water resources in an integrated manner requires the involvement of all those stakeholders who have an interest in its allocation, usage as well as wastewater disposal and reuse. This chapter does not discuss participation and partnerships in great detail as these issues are covered in greater detail in other chapters in the book (notably Chapter 11), but a brief description is included here in order to emphasis the importance for the successful implementation of stormwater management strategies. The interactions of different stakeholder groups with vested interests in water resources emphasise the importance of participation as a key instrument in the development of policy and plans that are acceptable to all stakeholders. Therefore, the involvement local stakeholders and consultation with communities is an integral component of the planning process.

As well as the agency (usually a municipality or other local authority), there is a wide range of stakeholder groups with interests in urban drainage (see Table 3.2).

Table 3.2 Stakeholder groups and their interests and priorities in urban drainage planning.

Stakeholder group	Interests and priorities
Public and community leaders	Largest community of residents and service users/beneficiaries
Slum dwellers	Low-income communities who often inhabit areas which are at risk from flood
Land developers	Construction of new developments for new housing or industry
Farmers	Peri-urban community with agricultural interests
Environmentalists	Protection of quality of water resources and conservation of natural habitat
Local politicians	Priority issues responding to local constituents' demands
Councillor and civil servants	Trade-off between cost–benefits in relation to municipal expenditure
Architects and land planners	Planning and design of urban space
Private sector, and the business community	Protection of industrial and commercial interests

Adapted from Bhattarai and Neupane 2001.

Through the involvement of local stakeholders in the planning process, the management of stormwater resources may be assigned to the most appropriate stakeholder at the lowest appropriate level. In theory, this should help to promote local *ownership* of the drainage plan and subsequently increases the responsibility for implementation.

IUWM requires the involvement of all stakeholders and it is important to consider their existing and potential future roles in stormwater management. However, getting stakeholders involved in IUWM is not without difficulties as there are many different interests from different sectors of society. As described in Chapter 11, conflicting and vested interests, a lack of incentives to co-operate and inadequate government support are some of the problems encountered in implementing participatory approaches to decision-making and planning.

In order to manage this process and to co-ordinate initiatives that arise in response to the planning process, there remains the need for local authorities to ensure that IUWM initiatives are implemented in a way that satisfies the demands from local stakeholders, whilst concurrently meeting obligations of government policies and legislation. This involves bringing together citizen groups, local agencies and states to work together on plans for the community and environmental improvements. Box 3.4 provides an example where community-based organisations (CBO), a local NGO, the water company and the municipality worked together as institutional partners to plan, design and implement an urban drainage infrastructure project in Dominican Republic.

Box 3.4 Participation and partnerships in urban environmental rehabilitation in Santo Domingo, Dominican Republic

An example of a successful slum-upgrading programme is the World Bank-supported pilot project where an NGO (*Insituto Dominicano de Desarrollo Integral*) has worked in partnership with various CBOs to construct storm drainage as part of an integrated approach towards urban upgrading and local environmental management. As well as drain construction, the project included paving and disaster-mitigating retaining walls, latrines, drinking water supply and collection of solid waste. A socio-economic and environmental survey of the entire settlement was undertaken with the community to help plan and prioritise investments.

Once the integrated environmental rehabilitation plan was in place, the NGO/CBO team encouraged the community to offer volunteer unskilled labour such as clearing garbage heaps and digging trenches. The team hired local masons to build sewerage and storm drainage along the bottom of the ravines and covered them with walkways in collaboration with the water company who installed piped water distribution to most homes. Concrete pedestrian paths and stairs with tracks for bicycles, carts and motorbikes over the storm drainage system up the side of gullies were also constructed. Grills over drains were installed to prevent solid waste from clogging the system and polluting the river downstream. A key disaster-mitigating infrastructure element of the project involved the installations of slope-stabilising concrete-reinforced walls that secure the terraces were people have built their homes.

Together, the CBO and NGO established a solid waste collection community micro-enterprise in order to ensure that the drainage system does not get blocked. While not fully self-sustaining, the programme is an excellent example of harnessing the energy of communities to help themselves. But ironically, the more affluent communities in the other parts of Santo Domingo do not have to pay any fees for municipal solid waste management, which highlights the need to address socio-economic and political inequalities as well as those related to infrastructure and services.

Source: Chavez 2002

3.6 REFERENCES

Amena, Z., Babel, M. S., Lacoursière, J. O., Vought, L. B-M. and Mark, O. (2004) Water quality modeling and management for the Nam Pasak channel in Vientiane, Lao PDR. *Proceedings of the 6th International Conference on Hydroinformatics*, Liong, Singapore, 21–24 June. (eds. Phoon and Babovic), World Scientific Publishing Company.

Bhattarai, S. and Neupane, B. (2001) Informed decision making for drainage management. *Water, Sanitation and Hygiene: Challenges of the Millennium, Proceedings of the 26th WEDC International Conference*, Dhaka, Bangladesh.

Chavez, R. (2002) Barrios El Café, La Mina and Hermanas Mirabal. In: *Santo Domingo: A Best Practice in Urban Environmental Rehabilitation*. World Bank Thematic Group on Services to the Urban Poor, Urban Notes on Upgrading Experiences No. 4 November 2002, The World Bank, Washington, DC.

CWP (1998) *Better Site Design: A Handbook for Changing Development Rules in Your Community*. Center for Watershed Protection, USA.

Diacon, D. (1997) *Slum Networking: An Innovative Approach to Urban Development*. Building and Social Housing Foundation, Coalville, Leicestershire, UK.

GWP (2000) *Integrated Water Resource Management*. Global Water Partnership – Technical Advisory Committee TAC Background Paper No. 4, Stockholm, Sweden.

Hashim, A. and Al-Mamun, A. (2002) Integrating urban stormwater management into drainage master plan of a tropical city. *International Conference on Urban Hydrology for the 21st Century*, 14–16 October, Kuala Lumpur, pp. 611–624.

ICWE (1992) *Dublin Statement on Water and Sustainable Development*. International Conference on Water and Environment, Dublin, 29–31 December. www.unesco.org/science/waterday2000/dublin.htm

Larsson, R., Vought, L. and Lacoursière, J. (2002) Urban drainage in Vientiane, Lao PDR management strategies for improved multipurpose use of an open channel system. *International Conference on Urban Hydrology for the 21st Century*, 14–16 October, Kuala Lumpur, pp. 599–610.

Parikh, H. H. (2001) Slum networking – using slums to save cities. In: *Frontiers in Urban Water Management* (eds. C. Maksimović and J. A. Tejada-Guibert), IWA Publications, London, UK, pp. 238–245.

Verma, G. D. (undated) *Indore's Slum Project – A Worm's Eye View from the Ground*. The Best Practices and Local Leadership Programme (BLP) Internet site. http://www.blpnet.org/blp/learning/casestudies

World Commission on Environment and Development (1987). *Our Common Future*. The Brundtland Commission Report, Oxford University Press, New York.

4
Policies and institutional frameworks

This chapter describes policy-related and institutional issues relating to the control of runoff and flooding in urban areas. Many of these issues relate to generic aspects of infrastructure and service provision in developing countries, including urban drainage, but they become especially important when considering integrated approaches towards stormwater management (see Chapter 3). Without clearly defined policies that are clearly structured in a way that each organisation understands and enacts upon their roles and responsibilities, it becomes inevitable that fragmented roles and responsibilities will contribute towards problems related to urban stormwater management as described in Chapter 1.

In order to achieve the challenges posed by integrated water resource management (IWRM), greater emphasis is required for the development of policies that encourage greater interactions and partnerships between different institutional stakeholders. In addition, to support the implementation of IWRM policies requires capacity building to promote organisational development and training of staff within these organisations.

4.1 POLICY FORMULATION

At one level, a policy is simply a general principle adopted by an individual or collective group of like-minded individuals, which expresses their opinions and

© 2005 IWA Publishing. *Urban Stormwater Management in Developing Countries* by Jonathan Parkinson and Ole Mark. ISBN: 1843390574. Published by IWA Publishing, London, UK.

how they aim to work together towards a common goal. At another level, it becomes a line of argument rationalising the course of action of a government, company, or NGO, which guides the organisation's overall orientation and management strategy.

National level policies formulated by central governmental agencies have far reaching implications for operational practices within the sector. These may subsequently form the basis for regulatory instruments, which provide the legal basis for the implementation of stormwater directives.

The type of policy dictates who is responsible for policy formulation and who is subsequently responsible for its implementation. Thus, existing practices and institutions are often the legacy of past policies combined with many years of working in a certain mode (Nielson 1998).

Policies are generally encapsulated in the form of written statements and official documents. These define the orientation of future activities and have both direct and indirect influences on current and future operational practices of institutions. Policies may lay down the details of institutional roles and responsibilities for policy implementation and may also include implementation guidelines such as regulatory procedures and technical standards.

In addition to policies that have a direct relevance to urban stormwater management practices, policies in other sectors may also influence the nature of flooding and urban runoff problems. For example, although policies related to urban development are generally not formulated from the perspective of stormwater management, they may indirectly affect the rainfall-runoff patterns and the distribution of flooding due to changes in impermeable areas.

4.1.1 Principles of policy

There are a number of prerequisites which affect the successful adoption of urban stormwater management policies. Firstly, it is important that they are understood and accepted. Without a general understanding and acceptance there is little likelihood that they will be successfully implemented. In addition to this, policies need to provide a clear designation of responsibilities for different stakeholders within a framework for action, which satisfy the following criteria:

- *Coherent*: defined by a set of logical and rational arguments that are expressed in a clear and consistent manner.
- *Congruous*: consistent with other policies that affect the nature of stormwater management problems.
- *Co-ordinated*: encouraging various stakeholders to act collectively and to form partnerships to maximise the potential benefits of their individual actions.
- *Comprehensive*: providing the basis for action at a wide range of levels to address a range of problems.
- *Certain*: enactment of policy needs to be predictable, transparent and clearly define the desired behaviour.

4.1.2 How policies are made?

The origin of any policy development at any level (national, institutional or individual) is a desire to change something. The next step in the process is the formulation of objectives in order to address the question of 'where do we go from here?' (Nielson 1998). At a basic level, successful policies are those whose constituents relate to as being sensible and for a common good. The key to policy-making is therefore to involve relevant stakeholders in the decision-making process in order to encourage ownership and support for the resultant policy.

In order to ensure that policies are accepted, there is a need to include a broad range of stakeholders from government agencies, companies, non-governmental organisations (NGOs) and community representatives in the process of policy development. This process should therefore be based upon principles of participation to ensure that the policy reflects the views of those who will be affected by its implementation. A democratic decision-making process provides individuals with an opportunity to express their views in relation to policy development.

Therefore, the policy must be translated and communicated in a way that is understandable to various stakeholders and it must be relevant to their concerns and interests; otherwise, it may be misunderstood and opposed. Communication is a key component in the development and subsequent implementation of policies in order to generate understanding, a broad acceptance and support for the policy framework and action plan.

4.1.3 The implications of policy implementation

Policies may have a direct influence on the institutional framework and the roles and responsibilities of different agencies and organisations. The choice of policy will direct whether the government takes on the role and responsibility of service provider and implementing agency, or enabler and regulator, in which case it does not involve itself directly with implementation. Depending upon how policies are formulated, they may also influence the modes of operation of these organisations through the introduction of directives, which introduce new or ban existing activities or working practices, or incentives which directly or indirectly encourage improved efficiency or operation. Box 4.1 describes the development of the Chilean Stormwater Act and the potential impacts that changes in policy may have upon the cost imposed upon beneficiaries to pay for urban drainage related services.

It is important that policies are developed in way to ensure that there are sufficient resources and technical capacity to implement them. It is unrealistic for low-income countries to adopt policies that they do not have the resources (or skills) to implement. Another important consideration relates to the mechanisms for policy monitoring and regulation (including enforcement) to ensure that the policy is being adhered to. These considerations support the argument that policies need to relate to the level of development.

> **Box 4.1 Implications of policy on financing and stormwater charges in Chile**
>
> After many years with frequent flooding in cities throughout the country, a new Stormwater Act was approved by the Congress in November 1997. The Act is targeted towards cities of 50,000 inhabitants or more and stipulates that each of these prepares an urban stormwater master plan describing the infrastructure and associated investment requirements. One of the most important articles of the Act makes it mandatory for developers to include stormwater infrastructure for all new urban developments. This inevitably increases the cost of the development, which will subsequently be passed onto the owners or leasers of the buildings.
>
> By the end of 2003, after stormwater master plans were prepared by most of the local authorities in the countries, the scale of additional investment that would be required for the main drainage networks became apparent. As a result, a proposal for modification of the 1997 Act has been discussed by Congress, which would have widespread implications on the financing of urban drainage infrastructure. The proposed modification involves including the stormwater charge as part of each householder's monthly water rates bill and a condition that if the bill is not paid, then the householder may have their water service cut. If the modification to the Act is approved, it is estimated that this could lead to an increase in water bills in some of the cities of Chile by as much as 40%. Although the bill includes the possibility of government subsidies to offset the extra charge, the increased charge is potentially a big problem, particularly for households without sufficient income to pay.
>
> *Source*: Professor Bonifacio L. Fernández
> Catholic University of Chile

4.2 POLICIES FOR RUNOFF CONTROL

Policies for runoff control refer to interventions and instruments to meet the objectives of stormwater management as defined in Chapter 3. These policies should ensure that stormwater management programmes are established and implemented for all major urban catchments. The following proposed recommendations may be adopted in order to address the most common stormwater management problems encountered in urban areas[1]:

Institutions

- Definition of explicit roles for different government agencies at different levels and separation of the institutional roles of stormwater manager and regulator to different utilities.

---

[1] These recommendations are based on a report by Australia's Commonwealth Scientific and Industrial Research Organisation. (*Source*: Introduction to Urban Stormwater Management in Australia http://www.deh.gov.au/coasts/publications/stormwater)

- Reform of the institutional structures for stormwater management to complement the utilities responsible for other aspects of urban water management (water supply, wastewater collection, solid waste management, etc.).

Finance

- Reform of drainage taxation systems so that charges more accurately reflect the costs of stormwater management programmes.
- Urban catchment agencies should have the power to obtain revenue either from local government or though direct charges on the private and public sectors, and to employ staff and commission works.

Environment

- Stormwater standards should be set within the context of national environmental quality objectives and the regulatory framework for pollution control.
- A water quality strategy should identify environmental values and complementary integrated catchment management activities.

Box 4.2 describes how some of these principles have been adopted in Malaysia as official government policy. Further details about the technologies promoted to meet this policy directive are found in Chapter 7.

Attempts to introduce pollution protection policies and environmental standards related to stormwater discharges in developing countries remains uncommon. However, evidence from industrialised countries demonstrates that as the quality of water resources deteriorates and the public awareness of environmental issues increases, there is increased pressure on governments to develop polices and mechanisms for pollution control.

In the past, the performance of sewer systems was characterised by the number of overflow events per year. This criterion was feasible to monitor and implement, but it did not account for the actual volumes spilled. Since the advent of relatively cheap and reliable urban drainage models for the calculation of overflow volumes and pollution emissions, the focus has changed from a frequency-based standard to a consideration of the yearly overflow volumes.

The total overflow volume over a period of time is particularly appropriate for discharges into lakes which are likely to suffer from chronic pollution problems where pollutants such as nitrogen, phosphorus and heavy metals accumulate over a period of time (see Chapter 1). On the other hand, the loads from extreme events causing high Ammonia or low-Oxygen concentrations may be the dominant processes in rivers and it will be this condition that dictates the standard adopted by the policy.

Policies for flood mitigation and risk management should recognise that it is impossible to protect all citizens at all times from the impacts of flooding. The policies should therefore be based on a realistic assessment of risk – both real and perceived. The emphasis of policies for flood management needs to shift from one of emergency response towards an emphasis on risk reduction. Building social assets can increase the chances of greater self-reliance amongst households and neighbourhoods (Sanderson 2002), but this requires some significant changes in policy, especially in relation to the low-income communities living in informal settlements.

> **Box 4.2 New approaches towards urban stormwater management in Malaysia**
>
> On 21st June 2000, the Malaysian government approved new legislation to replace the outdated Urban Drainage Design Standards and Procedures from 1975. The Department of Irrigation and Drainage (DID) now adopts this legislation, which specifies that all new developments must adopt technologies for source control of stormwater runoff. Overall, the policy has been well received, but there are issues that relate to the financial implications of the new policy that have yet to be fully resolved. There also remains the challenge to ensure that the administration of the planning, design and maintenance of stormwater management systems is consistent across the relevant local, state and federal authorities. In addition, it is important that urban planners, engineers and landscape architects adopt a unified approach within the planning and design of infrastructure in the urban environment. Therefore, to accompany the new legislation, a 'Stormwater Management Manual for Malaysia' has been produced to provide guidance to all regulators, planners and designers who are involved in stormwater management about the wide range of technologies that are available for source control. The manual also considers problems related to flooding, pollution, and soil erosion in order to promote good practices and disseminate information about various technologies. As described in Chapter 7, the *Universiti Sains Malaysia,* in collaboration with DID, has developed a demonstration project on the campus with the installation of swales, wetlands and various other sources control technologies to demonstrate the advantages of this new approach.
>
> *Source*: Zakaria *et al.* (2004)

4.3 POLICIES RELATED TO LAND USE

Policies for flood protection relate to the frequency of flooding of different types of land use. Stormwater drainage systems are conventionally designed to drain urban runoff from built-up areas so that flooding should occur no more frequently than once during a specified return period (as described in Chapter 6). Hence, it is necessary to design drainage systems with return period depending on the land use and classification of the urban area.

It is not possible to control urban runoff problems without considering the source of the problems related to the process of urbanisation as described in Chapter 1. Stormwater runoff problems related to flow and quality are intrinsically linked to increasing impermeable surfaces and land-based activities. Therefore, urban developments should be planned in a way so as to mitigate potential flood problems and control the use of land and land-based activities.

There is a need to develop policies and regulatory instruments to restrain uncontrolled development and place controls on certain types of urban developments in areas that are at high risk from flood. These regulations may specify the types of activities, defining the types of construction permissible and placing limits on the housing density. However, as the demand for land in cities is high, it is often not possible to assign land purely for flood management purposes. Therefore, land often needs to have an alternative use in order to ensure that informal settlements do not appear and, as discussed in Chapter 10, innovative approaches to land use control may be employed. These may be supported by taxation measures that provide financial incentives that guide development away from areas at high-risk from flooding.

Low-income communities, often live in flood-risk areas such as flood plains or riverbanks. Thus, integration of informal settlements within the city stormwater management plans is therefore a high priority. Urban drainage systems cannot be designed in isolation from the communities they serve and urban authorities should consider alternative approaches towards urban planning in which informal settlements and squatter communities are seen to be an integral part of the socio-economic and political part of the urban society – as in the example of slum networking described in Chapter 3.

In Brazil, there are increasing moves towards accepting informal settlements and the communities living there as part of the formal urban environment. In some cities (notably São Paulo) there have been progressive reforms towards the integration of favelas (slums) into the cities infrastructure and projects to provide services to these areas. The community-based watershed management (CBWM) in Santo André described in Box 4.3 aims to avoid problems arising from human

---

Box 4.3 Community-based watershed management (CBWM) in Santo André, Brazil

The CBWM project was implemented by the Municipality of Santo André in the State of São Paulo in collaboration with the Centre for Human Settlements–University of British Columbia and with funding from Canadian International Development Agency (CIDA). The project aimed to avoid problems arising from human settlement in environmentally sensitive areas around the Billings Reservoir, which provides much of São Paulo's water supply. The overriding problem related to illegality of tenure and the fact that the households could not receive any infrastructure, as the existing law would forbid it. The CBWM project helped residents regarding land tenure issues and encouraged amendments in the law to allow for provision of infrastructure and services to these areas. In addition to the political commitment, the project required new operational tools for environmental management that move beyond traditional top-down master planning that have relied upon a restrictive legalistic approach towards environmental management. The CBWM process involves the inhabitants of illegal settlements and other stakeholders in an adaptive planning framework, which aims to gain consensus on the objectives prior to the development and implementation of the plan. It was also very important to relocate people who had their houses near streams, close to the reservoir or in dangerous areas. In doing so, a design for the whole settlement to improve stormwater management was proposed and discussed with the community. This included a new street design for stormwater drainage and paving with increased permeability.

Source: van Horen 2001

settlement in environmentally sensitive areas and to integrate informal settlement areas more effectively into the urban system.

Resettlement of households who have constructed dwellings on flood plains or slopes at high-risk from landslides is also a priority. It is important to consider a policy towards resettlement of communities already established and living in risk areas. However, policies that propose resettlement for communities occupying land illegally can be extremely contentious, especially where the communities have been in occupation for many years. Although these communities may suffer from the impacts of flooding on a regular basis, there are usually good reasons why they live there (see Chapter 2). Thus, proposals to relocate them to other areas that are a long way away from the areas of employment and other urban services, will not be well-received without due consideration of the needs of the communities to support their livelihoods.

4.4 INSTITUTIONAL FRAMEWORKS FOR POLICY IMPLEMENTATION

In the majority of situations, it is the responsibility of local government and municipal agencies to develop and implement urban runoff and flood control strategies. The main decisions related to urban stormwater management are made by local governmental institutions and water-related companies, whereas for regional issues, federal government and ministries take over the full responsibility (Andjelkovic 2001). According to Fox (1994), the roles of different institutions and the level at which they operate has considerable influence on the provision of urban services as follows:

- *At the national level*: an agency can be responsible for long-term planning, standard-setting, finance, procurement of imported parts and technical assistance, co-ordination of training, and provision of advice and support.
- *At the regional level*: an agency can be created with responsibility for the development and finance of nationally approved standards and regulations, supervision and support to local systems, and planning and training for local management and technical staff.
- *At the local level*: entities can be developed with responsibility for management, collection of fees and monitoring of use, operation and maintenance.

In some situations, local authorities may define their own policies but they are often influenced heavily by central government and national agencies, who often maintain control of the total budget and expenditure as shown in Figure 4.1. In addition, governments may have entirely different ministries responsible for emergency management and urban development with little knowledge of each other's activities.

Local governments may become motivated to act on drainage issues when flooding affects the business district, as in Cabanatuan in the Philippines, where the local business community put pressure on the mayor to invest in drainage infrastructure. In Kampala in Uganda the local authorities had neglected for years

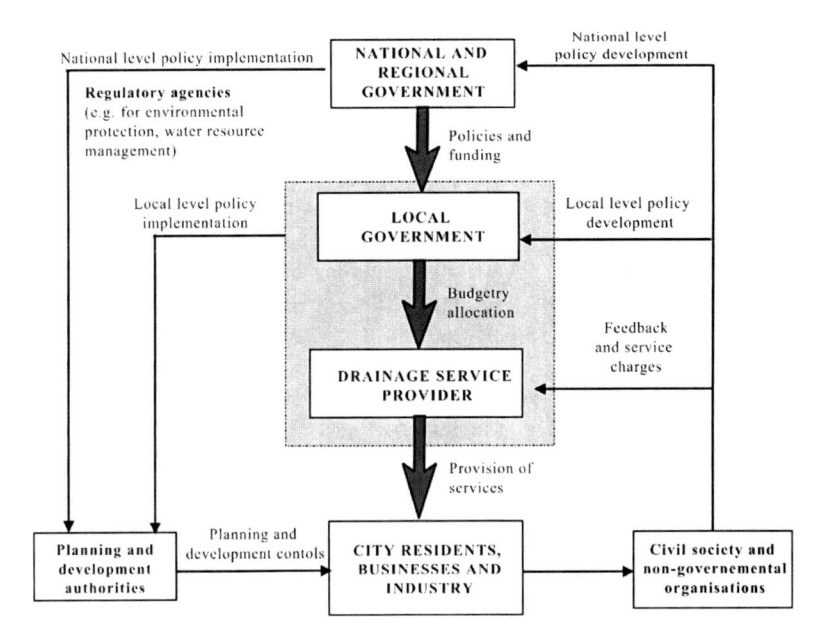

Figure 4.1 Interactions between stakeholders in integrated urban stormwater management.

to protect past investments in the Nakivubo channel from settlement encroachment and obstruction with solid waste. There, as well as in Ethiopia, recent reforms expanding local democracy have raised the profile of drainage as a priority for public expenditure (World Development Report 2003) and in Brazil, the introduction of participatory budgeting shows a similar trend (see Chapter 12).

Given the broad set of objectives of Integrated Urban Water Management (IUWM) described in Chapter 3, it is unrealistic to assume that one organisation will be able to assume responsibility for all of these. Therefore it will be necessary to have a diverse set of institutional responsibility assigned to different organisations (e.g. urban planning authority, and agencies responsible for roads and paving, sewerage and solid waste management). However, it is important that one organisation is seen to take the lead role and takes the ultimate responsibility for programme management and co-ordination.

This is especially important when considering the roles and responsibilities in decentralised municipal management arrangements for urban stormwater management. These can be fairly complex, especially in larger cities where it is necessary to have a sub-division of areas as described below in Box 4.4. There remains a need for a centralised agency to maintain overall responsibility, co-ordination management for stormwater infrastructure and related services for operation and maintenance. A centralised agency therefore needs to be involved in order to ensure that each small locality does not resolve its drainage problem by exacerbating that of its neighbours (Kolsky 1999).

> **Box 4.4 Decentralised municipal management arrangements for urban stormwater management in Kampala, Uganda**
>
> Larger cities are typically divided into more than one geographical area or zone in which different divisions are responsible for urban management. The institutional arrangements for managing the infrastructure and services associated with drainage system operation and maintenance in Kampala are decentralised, but the directives and overall policy remain centralised – the functions of planning and capital development. The technical arm of the City Council is the City Engineer and surveyors department, which is responsible for planning, design, and operation and maintenance of the city infrastructure. The Deputy City Engineer and surveyor have overall responsibility for operations and maintenance. In each division, there is an executive engineer in charge of highways and drainage who reports to the senior executive engineer (operations) in charge of the division. In turn, he reports to the Principal Executive Engineer (Highways maintenance). The overall tasks of these divisions are to maintain the roads and drainage.
>
> *Source*: Rugumayo 1999

4.5 INSTITUTIONAL DEVELOPMENT AND ORGANISATIONAL STRENGTHENING

In developing countries, the institutional systems for effective implementation of land planning and development controls are often weak. Just having policies on paper is no guarantee of an effective programme. Institutions must be organised in such a way as to implement the policies and carry out the procedures. Institutional factors have a significant impact on the successful implementation of urban runoff control policies and this is linked directly to the efficacy of urban management. Therefore, prevention and control of urban runoff requires well-defined policies and procedures, as well as effective programmes and institutional structures.

To achieve this, the responsibility of each agency involved in an urban runoff control project should be made very clear and the co-ordination between these different organisations is of crucial importance. Excessive adherence to complex procedures combined with the bureaucratic nature of government agencies in some developing countries can be an important factor in the inefficient operation of the organisations. It must be accompanied by a real institutional commitment to change ineffective and outmoded structures, to break through political and bureaucratic constraints. Successful policies need to seek ways in which this may be achieved and the following ideas may provide ways to overcome these institutional barriers (Losco 1994):

(1) *Review relevant policies and programmes directly or indirectly related to stormwater management practices and assess their effectiveness*: this may involve looking at a range of stormwater and water quality programmes scattered across different agencies and departments.

(2) *Determine the motivations and goals of the key agencies*: they may be acting in response to flooding, environment health or pollution problem, but although there may be crossover, the agencies may not understand the objectives of the others. In addition, not all organisations will be subject to the same level of political pressure.
(3) *Determine whether political support exists*: this should include both official political representation and NGOs to assess the community awareness about issues related to stormwater management.
(4) *Identify funding options*: this might include a feasibility study of different funding sources such as a stormwater utility and development of a fee structure (see Chapter 12).
(5) *Consider the limitations of available technology*: the potential for solving a problem may be limited by many factors over which the implementing authority has no control. This includes performance limitations of technology as well as site-specific constraints.

Technical problems are intrinsically linked to management and institutional problems. Often a deficiency in an easily identifiable area of urban drainage is identified as the primary source problem, when in reality the deficiency identified is a symptom of a larger problem related to institutional performance (DFID 2003). Therefore, there is a danger of misdiagnosing problems by focussing too much on technical problems, which ignore more fundamental institutional deficiencies and the importance of management leadership to deal with the problems. Most efforts in the past have paid insufficient attention to institutional problem diagnosis. Therefore, the process of problem identification and diagnosis requires a fundamentally different approach than those that focus solely on technical issues.

Institutional development is of crucial importance and a prerequisite where changes to policies are introduced. It is also necessary to ensure that institutional capacity matches its technical ability to deal with urban runoff problems. Improved institutional capacity implies greater capacity for organisation and technical competence. Training and organisational strengthening are complimentary components of a capacity-building strategy in combination with supporting policies which create an enabling environment for the development of effective organisations (as illustrated in Figure 4.2).

Whereas institutional development refers to the overall structure for urban stormwater management, organisational strengthening is all about getting things to work within that structure. Cullivan *et al.* (1998) provide a set of procedures to assist in the diagnosis of institutional deficiencies in the water and wastewater sector according to each major category of organisational function related to the following:

(1) general management;
(2) human resources management;
(3) financial and commercial;
(4) planning, design and construction;
(5) operation and maintenance.

62   Urban stormwater management in developing countries

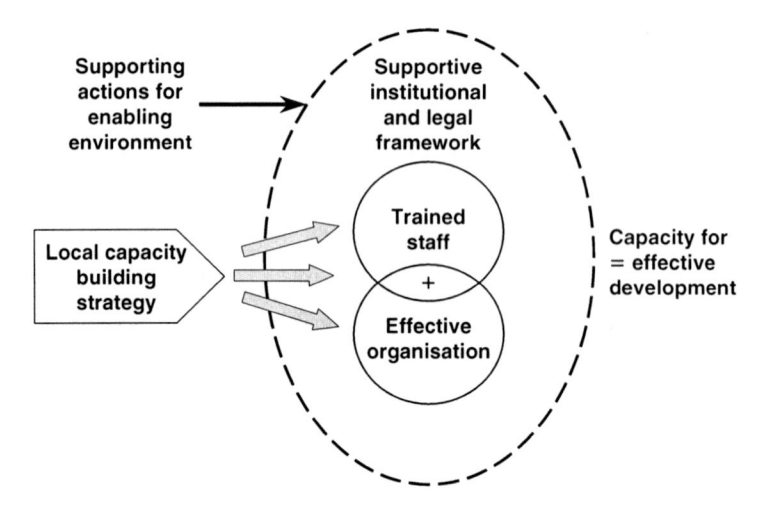

Figure 4.2 Concept of capacity building (Peltenburg *et al.* 1996).

Based upon this, various recommendations for organisation strengthening and human resource development for a stormwater management utility are shown below in Table 4.1.

Table 4.1 Recommendations for organisational strengthening for a stormwater management utility. (Based on Cullivan *et al.* 1998.)

Category	Recommendation
Management and administration	Improve management information systems Develop personnel policies manual
Financing	Undertake studies to derive cost-benefit relationships Develop billing and accounting procedures
Technical capability	Training system for the development of management skills and improve technical skills
Technical procedures	Formalise design procedures and standardise stormwater designs
Interactions with key external institutions	Organise meetings with other interested stakeholders to share knowledge and experience and to provide an informal opportunity for collaboration
Community liaison	Organise community meetings and discuss issues with community-based organisations

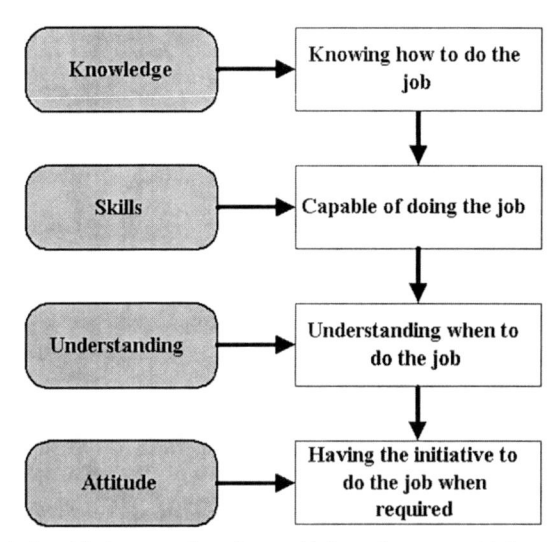

Figure 4.3 Relationship between learning and job performance. (Adapted from Rashid and Shamsuddin 2004.)

4.5.1 Training and human resource development

The effectiveness of an organisation depends on complex behavioural factors related to individuals just as much as the structure of the organisation itself. Figure 4.3 shows the relationship between learning and job performance. In consideration of the incentives within organisations, it is necessary to consider the structure, the organisation and the staffing. Any organisation will only work well if the staff, working at all levels, have incentives; these should cover incentives for each unit, as well as individual worker objectives. Each position in the organisation needs to be valued and rewarded according to performance. In a situation where only higher levels of management and technical staff are rewarded this will act as disincentives to the rest of the workforce and create an unhealthy working environment.

A training needs assessment process begins when a decision is made by the management to sanction the use of systematic needs assessment in locating appropriate targets for training. A comparison of methods for assessing training needs is found in Table 4.2. If training needs assessment is new to the organisation, it may be necessary to appoint and train staff or to engage competent assistance from outside. Strong management support is required to give credibility to assessment activities in the eyes of the organisational units affected.

Capacity-building activities will vary from overall improvements in the technical and managerial capacity of staffing to the formulation of procedures that promote accountability and transparency, and to the introduction of information technologies to assist in administrative functions. The establishment of knowledge networks, to enhance learning and institutional memory that can be shared by the public, the private sector, and community agents should be part of capacity-building strategies to promote and support policy.

Table 4.2 Comparison of methods for assessing training needs (Tees *et al.* 1992).

Conventional	Systematic
Is carried out for the agency by training institutions using standardised methods	Is carried out by the agency to identify its own solutions for gaps in performance
Focuses on a single source of data, the person who responds to a survey request	Focuses on multiple data sources to verify training solutions for performance problems
Targets levels of the agency where there are needs corresponding to the capabilities of the training institution	Targets various levels of the agency depending on where problems or changing situations are found
Centres on subject matter for training and the agency's reaction to a list of topics	Centres on concrete problems and the consequences of planned changes for performance in the organisation
Depends on the skill of outside training institutions to carry out the assessment	Depends on management commitment and personnel who are able to carry out needs assessment on a regular basis
Does not distinguish between training-related needs and needs for organisational improvement	Distinguishes between training needs and non-training needs and provides linkages between the two types of needs

4.6 REFERENCES

Andjelkovic, I. (2001) *Guidelines on Non-structural Measures in Urban Flood Management.* IHP-V Technical Documents in Hydrology. No. 50. Project IHP-V Project 7.UNESCO, International Hydrological Programme, Paris, France.

Cullivan, D., Tippett, B., Edwards, D.B., Rosensweig, F. and McCaffery, J. (1988) *Guidelines for Institutional Assessment Water and Wastewater Institutions.* WASH Technical Report No. 37. Water and Sanitation for Health Project. US Agency for International Development, Washington DC, USA.

DFID (2003) *Promoting Institutional and Organisational Development.* Department for International Development, London, UK.

Fox, W. (1994) *Strategic Options for Urban Infrastructure Management.* UMP Policy Paper No. 17, Urban Management Program, World Bank, Washington DC, USA.

Kolsky, P. (1999) *Performance-based Evaluation of Surface-Water Drainage for Low-Income Communities: A Case Study in Indore, Madhya Pradesh.* PhD thesis, London School of Hygiene and Tropical Medicine, University of London, London.

Nielson, T.K. (1998) *The Technological Upgrading of Service Institutions.* Intermediate Technology Publications Ltd., London, UK.

Peltenburg, M., Davidson, F., Teerlink, H. and Wakely, P. (1996) *Building Capacity for Better Cities-Concepts and Strategies.* Institute of Housing and Development Studies, Rotterdam, The Netherlands.

Rashid, H.Ur., and Shamsuddin, Sk.A.J. (2004) *Training Skill Development.* ITN-Bangladesh, Bangladesh University of Engineering and Technology, Dhaka, Bangladesh.

Rugumayo, A. (1999) *Operations and Maintenance: Kampala City Council – A Case Study.* Short Course in Technology, Management and Operations of Urban Drainage Systems in Africa – The Present and the Future. 28th March–3rd April. Kampala, Ghana.

Sanderson, D. (2000) Cities, disasters and livelihoods. *Environment and Urbanization* **12**(2), 93–102. International Institute for Environment and Development, London, UK.

Tees, D.W., You, N. and Fisher, F. (1992) *Manual for Training Needs Assessment in Human Settlements Organizations: A Systematic Approach to Assessing Training Needs*. United Nations Centre for Human Settlements (Habitat), HS/114/87/E, Nairobi, Kenya.

World Development Report (2003) Sustainable development in a dynamic world. Chapter 6 In *Getting the best from cities*. World Bank, Washington DC, USA.

Zakaria, N.A., Ab. Ghani A., Abdullah, R., Sidek, L.M., Kassim, A.H. and Ainan, A. (2004) MSMA – A new urban stormwater management manual for Malaysia. Proceedings of the 6th International Conference on Hydroscience and Engineering (ZCHE – 2004), Brisbane, Australia 30 May–3 June.

5
Planning and assessment of improvement options

In order to approach the complex set of issues involved in the design and implementation of urban drainage systems, especially taking into account multiple objectives of integrated urban water management (IUWM), it is essential to plan effectively. Planning is therefore one of the most important management functions for institutions responsible for the design and implementation of urban runoff and flood control strategies. In order to ensure that the solutions derived by the planning process respond to real needs, it is important to involve local stakeholders in the assessment of problems and the formulation of solutions. This chapter deals with the various activities involved in the development of stormwater plans, including the data requirements for planning approaches towards data collection and how information management and decision-support systems can be used to assist the planning process.

5.1 WHAT IS PLANNING?

Planning is a term that is used to describe a diverse set of activities with the underlying objective to assess current problems and derive solutions to overcome these problems. Although planning is involved in many daily activities, at the more formal

© 2005 IWA Publishing. *Urban Stormwater Management in Developing Countries* by Jonathan Parkinson and Ole Mark. ISBN: 1843390574. Published by IWA Publishing, London, UK.

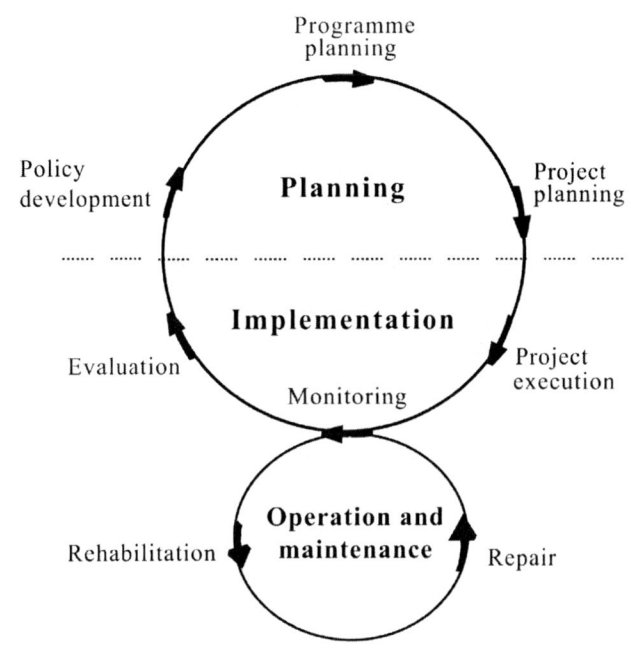

Figure 5.1 Relationship of different 'levels' of planning with the implementation, and operation and maintenance cycles.

level, the output from the planning process generally involves the production of a written document, which provides a record of the decisions made during the planning process and a framework to guide future activities and investments. Thus, the ultimate aim of the planning process is to generate:

(1) *Remedial* solutions, which focus on rectifying existing problems in the short term.
(2) *Preventive* solutions, which focus on problem analysis and the development of solutions to the causes of the problems with a longer-term perspective.

As shown in Figure 5.1, the planning process can focus on one of three levels relating to policy development, programme and project planning. Although the basic principles of planning are the same, the focus and nature of the activities involved in these planning activities will not be the same and involve different stakeholders.

Figure 5.1 also shows how monitoring and evaluation of operational performance should influence policy development, which in turn will orientate the development of future programmes and projects. It also indicates the importance of operation and maintenance as an integral part of the project implementation cycle and the fact that the identification of remedial and rehabilitation measures should be based on the results from a programme of monitoring of system performance.

Policy development

The definition of the policy objectives (see Chapter 4) involves a process of strategic planning, which defines the overall goals and long-term objectives related to the level of performance required. It should assign roles and institutional responsibilities for stormwater management and consider who will be responsible for regulation of standards. It should also define indicators for monitoring and evaluation, using performance and process indicators as a means of assessment of operational performance (see Chapter 9).

Programme planning

Once long-term strategic policy objectives have been defined, it is necessary to orientate these into a more defined programme of activities. Generally, the programme plan contains spatial and temporal elements, indicating where and when interventions should be initiated and who should be responsible for implementing these activities. These plans are often referred to as *master plans*, which define the human, technical and financial resources for implementation of specific initiatives and projects.

Project planning

The next stage of the planning process involves the preparation of project plans orientated towards achieving specific goals within given constraints of time and resources. Project planning involves a detailed assessment of local problems in specific areas, evaluation of potential solutions, preliminary design and consultation with local stakeholders to discuss the details of the proposed intervention.

5.2 THE PLANNING PROCESS

Although the planning process may be defined in terms of a greater number of steps, there are essentially three main stages as illustrated in Figure 5.2. Each of these three stages involves an iterative process. For instance, in the first stage, it will be necessary to collect and analyse data as part of the problem identification process. However, once problems have been identified, it may subsequently be necessary to collect more data in order to define these problems more clearly.

Similar to the other two stages, which also involve iterative processes – after the initial evaluation of the problems and formulation of solutions, it may be necessary to review the problems in greater detail. Likewise, the evaluation and comparison of solutions are inter-related activities, which are required to identify the most appropriate solutions.

Different approaches towards planning require different types of information and therefore the collection and analysis of information should be closely linked to the planning process. As data is often lacking, it will often be necessary to

Planning and assessment of improvement options

```
Decision to embark upon
    planning process
          ↓
  Information collection
  Problem identification
          ↓
   Assessment of problems
   Formulation of solutions
          ↓
   Evaluation of solutions
   Comparison of solutions
          ↓
   Selection of most
   appropriate solution
```

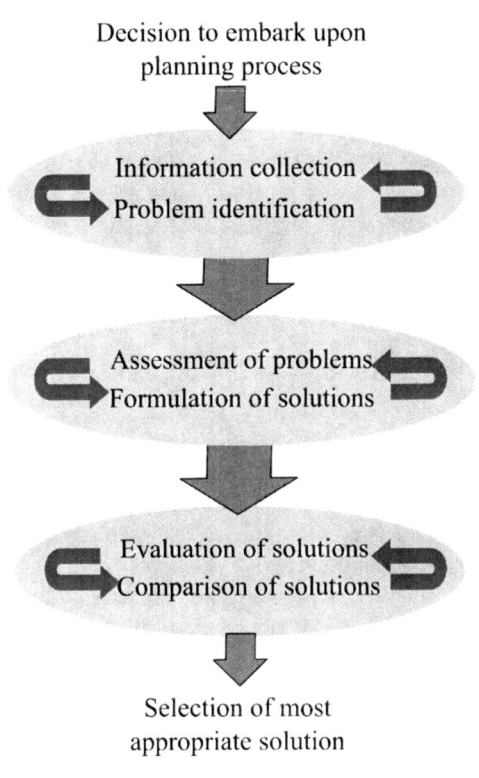

Figure 5.2 Stages in the planning process.

undertake surveys to gather information about existing drainage systems, areas of development and locations where flooding is recognised to be a problem (see Section 5.5).

It is very rare that there will be one unique and perfect solution to urban drainage and flooding problems. Normally, there will be a range of potential solutions and these will be dependent on the priorities and concerns of the *stakeholders*; that is, those people who have an interest in the outcome or will be affected by the implementation of the plan. As a result, planning is a complex process involving many individuals, groups or organisations, many of whom will have different perceptions about prevailing problems, motives to be involved in decision-making, as well as ideas about the sort of solutions that they think will be the most effective.

In some situations, these perceptions and opinions may be radically different and there will be a need to resolve these differences and find a common basis of understanding and agreement before proceeding further towards the development of the details of the plan. The final plan will often require trade-offs between competing goals related to various quantitative and qualitative aspects of urban stormwater management.

The way in which decisions are made during the planning process will also influence the way in which the plan is implemented. Thus, the approach adopted towards the planning process will have an important bearing on the resultant outcome; that is, the plan itself. Therefore, although the plan itself is important, in some respects, of equal, if not of greater importance is the *process* that is adopted during the planning activities. In the first instance, the planning process may simply be a way of engaging with the various stakeholders and a means to focus their attention to the issues of urban drainage and the need to develop solutions to problems related to flooding. As described below, the first step in the planning process itself should be a review of the existing situation.

5.3 REVIEW OF THE EXISTING SITUATION

In order to ensure that the formulation of solutions is not based on too many preconceived ideas about what the existing problems are, it will be important that the planning process starts with a realistic assessment of the existing situation and the availability of resources (Tayler *et al.* 2003). This involves a thorough exercise in data collection and collation, which should form the basis for problem identification, formulation of solutions and to evaluate the potential benefits of different improvement options. The range of information that is relevant to the planning and design of urban stormwater systems is both extensive, but in principle should aim to respond to the following questions.

What are the main physical and hydrological features of the catchment?
One of the initial activities will be to define the boundaries of the catchment and to collect physical and topographical data describing soil types, natural drainage paths and location of the watercourses into which urban runoff discharges. In addition, data describing land use and housing density are required to estimate the proportion of impermeable areas for calculating runoff. Various methods of data acquisition related to spatial and topographical data and data describing land use are described in Section 5.6.

What is the coverage of urban drainage infrastructure?
Often the level of detailed information describing the coverage and capacity of the existing drainage infrastructure is limited. Maps showing the location of constructed drains are often out of date and only partial. One of the first activities will therefore be to identify and survey routes of both natural and constructed drains.

What is the capacity of the urban drainage system?
Construction details may be available for some of the large drains in order to be able to make an initial assessment of the hydraulic capacity of the drainage system. However, it is important to consider that the theoretical capacity of the system according to the design will not be the same as the operational capacity. The actual capacity will probably be reduced significantly by the ingress of sediment

and refuse (see Chapter 9) and it may be necessary to carry out a survey to assess the levels of solid wastes in the drains and locations of persistent blockages and obstructions (Kolsky 1998).

What is the current status of flood conditions and water quality in receiving waters?
An evaluation of the scale and nature of problems related to flooding, drainage and contribution of urban runoff to pollution problems will be required in order to provide the basis for the formulation of potential solutions. This may involve an analysis of hydrological and hydraulic data to assist in the identification of flood risk areas.

What are the expectations of local stakeholders?
Participatory planning techniques may be used to assess who is affected and how problems are perceived, their level of expectation for improvement and the demand for improved services (see Chapter 12).

What is the availability of financial resources to invest in interventions?
The availability of financial resources will obviously be crucial for the implementation of any proposed intervention. A financial assessment should also involve an audit of current assets and an evaluation of previous flood damage. The assessment should also consider operational and maintenance costs and sources of revenue for cost recovery (see Chapter 12).

What policies or legislation will affect the development or implementation of proposed solutions to urban stormwater problems?
It will be important to review both current and proposed policies and legislation that may have a direct impact (e.g. frequency of flooding or water quality standards) or an indirect influence (e.g. urban development regulations) on the generation of runoff and its management.

What is the capacity of institutions to implement plans?
It will also be important to consider the existing institutional arrangements and how these relate to the management of existing infrastructure. The outcome from this might influence the success or otherwise of proposed solutions and different options for intervention. This should include an assessment of human resources and capacity of the agencies and local institutions that will be responsible for implementation of the plan and subsequently for operation and maintenance (as described in Chapter 4).

5.3.1 Problem analyses and formulation of solutions

There are a number of approaches towards problem analysis and the formulation of solutions, which may lead towards the identification of proposals to solve prevailing flood and urban drainage problems in the catchment. These are based on either a quantitative or a qualitative evaluation (or a mixture of the two) using results from the review of the existing situation described above.

The relationship between the qualitative and quantitative assessment is important and generally the best approach is to consider both. The key issue for those responsible for the planning is the extent to which it is necessary and possible to go on to assess these problems and possible responses to them in a more quantitative way. There is no one correct way of undertaking this stage in the planning process, but a number of different approaches, and each of which has validity.

Often there will have been previous attempts to plan and develop project proposals, but these may never have been implemented or may only have been implemented in part, often due to a lack of financing. Although the proposed solutions in these plans may be overly ambitious, they often contain good analysis of drainage problems (generally those related to technical issues) and may also provide indications as to the solutions that may be required. On the whole, these are likely to be more focused on structural remedial problems and less upon non-structural mitigation strategies or preventative solutions.

It is important to note that many problems may be solved by improvements in the operation and maintenance of the existing system and these may require very little in the way of investments in new infrastructure. In this situation it will be important to seek the root causes for problems which may result in non-structural solutions or solutions which are related to institutional issues. Of particular importance is the fact that the operational performance of the systems may be significantly impaired by the quantities of solid waste in the drains. Large-scale investments may result in only minor improvements to flooding if new constructions become clogged with sediment and other solid wastes. Where solid waste is identified to be a problem, an improved maintenance strategy should be considered (see Chapter 9).

It will be important to talk to both engineers and operators who know the existing system well and who have a good practical knowledge of the way it operates and its deficiencies. Drainage technicians may also remember previous attempts to implement solutions to drainage problems that failed and their knowledge of these may help ensure that the same mistakes are not repeated.

Another important approach towards the formulation of solutions is to talk to local community members, both individuals and groups, in order to discuss with them about what they think are the main causes of the problems and what they consider to be the best ways to solve these problem. Although these people may not have a good technical understanding, they may have a good intuition about the cause of problems and also have some good suggestions about how to solve them. This can be linked directly to the assessment of the initial situation using a skilled facilitator using participatory methodologies (see Chapter 11).

In addition to qualitative problem evaluation, it is important to undertake some quantitative assessment. This may be particularly useful to provide a more judgement-based analysis, which provides the basis for discussion and evaluation. Although it will be too early to do undertake any detailed computer simulations, it may be possible to use a model to undertake some quick initial assessments of various proposed improvement options.

Once these ideas have been collated, a few options are likely to emerge as candidates for the most appropriate solutions. The next step will be to evaluate and compare these according to further quantitative and qualitative criteria before making a final decision about which proposal is considered to be the most appropriate and approved for implementation.

5.4 EVALUATION AND COMPARISON OF ALTERNATIVE SOLUTIONS

The proposed solutions identified in the planning stage described above will need to be evaluated according to various criteria (such as those detailed below in the case study from Biratnagar in Nepal), which indicate their relative merits and potentially negative impacts. Based on the outcome from this, various alternatives can be compared in order to subsequently identify which option offers the most cost-effective and sustainable solution.

As in the previous stages in the planning process, there are two basic approaches involved in the evaluation and comparison of alternatives:

(1) *Quantitative assessment*, which defines the assessment in numerical terms; for example, the reduction in the frequency of floods events or the volume of stormwater discharges into receiving waters.
(2) *Qualitative assessment*, which takes into account perceptions of local stakeholders and relates to the quality of the service.

The challenge at this stage is to evaluate various proposed solutions from a qualitative as well as a quantitative perspective. Participatory planning needs to be complemented by some rigorous technical evaluation and comparison of options.

5.4.1 Quantitative assessment

As mentioned above, a computer model may be used in the process of the initial assessment of solutions. It is during the process of quantitative assessment where the application of computer models to simulate various solutions becomes particularly beneficial. As mentioned in Chapter 8, it is important to choose the right tool for the purpose and the development of a detailed computer model for intensive simulations may not always be necessary of appropriate. Due to the complexities of model construction and high data requirements, there may be a need to consider simpler tools, which enable a large number of scenarios to be evaluated without the need for a detailed modelling. One example of an approach, which uses a simple computer model for this purpose, is described in Box 5.1.

However, it is important to ensure that the data collected and knowledge base are sufficiently detailed and accurate before embarking on any planning activity involving computer models. Otherwise there is a danger that the construction of

> **Box 5.1 Urban Pollution Management (UPM) methodology**
>
> The UPM methodology has been developed as a planning approach in the UK for the management of urban wastewater discharges during wet weather. The aim of the methodology is to enable wastewater engineers and catchment planners to identify environmentally acceptable, cost-effective solutions to urban wastewater pollution problems and in particular to problems related to urban runoff. The methodology provides guidance and potential environmental criteria to be applied in the development and monitoring of strategies that are designed to protect receiving waters from chemical, bacteriological and aesthetic pollution caused by discharges from combined sewer overflows (CSOs). The UPM methodology is underpinned by wet weather environmental standards, with an appropriate modelling process being used to demonstrate compliance of the proposed scheme with these standards.
>
> One of the main concepts adopted by the UPM methodology is the need to consider the wastewater system (comprising the sewer system, the treatment plant and the receiving water) as a single entity in which a change to one part has implications for the other parts. The SIMPOL model produced by the Water Research Centre in the UK has been developed as a tool to examine the impact of wastewater discharges and assess the impact of various proposed improvement strategies. By integrating the wastewater and river systems into one model it becomes possible to examine the combined effects of all discharges and to track any knock-on effects further down the river system.
>
> This capability was particularly helpful in examining overall solutions for improving the middle and lower reaches of the River Tame in the south-west of the UK. The adoption of a standardised methodology helped in obtaining agreement with the Environment Agency for the need to balance investments in improved CSO structures with other improvements related to management of diffuse pollution from other sources of surface runoff, as well as from wastewater treatment plant effluents. The results from the study were also used by Severn Trent Water to obtain Environment Agency approval to commence detailed engineering design for specific CSO improvement schemes and to secure future funding for a drainage improvement programme up until 2010.
>
> *Source*: FWR 1998; Williams *et al.* 2003

the model will detract the attention away from some of the other more important issues, which cannot be included in the model.

5.4.2 Qualitative assessment

Qualitative assessment is a judgement-based process and involves consultation with key stakeholders and representatives from various stakeholder groups as part of the process of evaluation and comparison of various proposed solutions. The

Planning and assessment of improvement options 75

qualitative assessment may be based on participatory assessments, which means that those responsible for the planning process need to engage with and involve a wide range of local stakeholders and organisations with interests in flood mitigation and pollution control.

As described in Chapter 3, participation is a critical component of IUWM and participatory planning activities may also promote public awareness of urban drainage problems and promote local acceptance of the plan. More details of issues related to participation in the planning process are discussed in Chapter 11. However, participatory assessment involving communities should not be undertaken unless it is very clear how it is to be managed, by whom and what the objectives are. In addition, qualitative assessment should not only be focused on the public consultation. The consultation of specialists needs also to be used, especially where relevant information is not available, and the judgements and opinions of these experts may be critical in the process of evaluation of different solutions, as well as in the analysis of problems.

The final stage of the planning process is to combine qualitative and quantitative assessment in order to make a final assessment of the various options. Comparison of various options based on their physical, social and economic feasibility involves a complex process of assessment of value judgements and these factors cannot be approached in a linear approach; that is, by summation of the individual factors to give an aggregate value.

5.4.3 Decision-support systems in the planning process

An increasing number of decision-support systems are available to enable the user to consider a wide range of permutations, which incorporate a wider range of quantitative and qualitative factors that would otherwise be feasible (e.g. Fletcher *et al.* 2001; Abrishamchi and Tajrishi 2002). However it is important to emphasise that these may only be used to assist in the process of decision-making and not to make decisions.

Multi-criteria decision-making analysis (MCDA) is a methodology that provides a systematic framework for the comparison of various alternatives based on various quantitative and qualitative criteria. In order to aggregate the scores each criterion needs to be given a weighting according to the relative importance as perceived by local stakeholders and the significance they place on different issues. The process used to derive the weighting factors is based on consultation and participatory planning in which local stakeholders are asked to make judgements on their preference based on various objective and subjective factors related to different quantitative and qualitative criteria. This is inevitably a subjective process and caution needs to be places in the interpretation of the results. In MCDA procedures such as the Analytical Hierarchy Process, the aggregate impact of a decision is expressed as a numerical score, with a higher score indicating aggregate positive impact of the decision. As described below, this approach was used to analyse a set of objective functions in order to assess various proposals for improvements to the urban drainage system in Biratnagar, Nepal.

Case study: Consultation and stakeholder analysis in Biratnagar, Nepal

As a response to the scale of the urban drainage problems and the lack of resources to implement a drainage master plan prepared in 1989, a new approach was conceived during the development of a new plan for the municipality of Biratnagar in Nepal (Bhattarai and Neupane 2001). Three options for drainage improvement (A–C) were initially screened and subsequently appraised through a process of stakeholder consensus. The process of informed decision-making involving community members and other stakeholder groups required an evaluation of socio-economic, technical, environmental, managerial and financial factors (see Table 5.1).

Table 5.1   Factors for consideration in the assessment of drainage options.

Factor	Description	Objective function
Socio-economic		
Productivity	Output from the investment/input made by municipality and users	Maximise
Health	Health benefits to the people	Maximise
Affordability	Ability of the people to pay for the services, especially those related to operation and maintenance	Maximise
Agriculture	Adverse impact on farming practices	Minimise
Technical		
Risks	Technical risks and related uncertainties	Minimise
Technical capacity	Technical capability of the municipality (manpower, skills and availability of equipment)	Optimal use
Time	Time required to complete the project	Minimise
Environmental		
Social	Perceived adverse social impacts of the project	Minimise
Biological	Adverse impact on the environment	Minimise
Physical	Adverse impact on the physical environment	Minimise
Managerial		
Legal	Ability of each option to satisfy various legal requirements	Minimise
Institutional capability	Capability of the municipality as an institution	Optimal use
Stakeholder considerations	Conflict with other service delivery agencies (water supply, irrigation, roads, telecommunications)	Minimise
Financial		
Construction	Initial cost of construction	Minimise
Operation and maintenance	Operation and maintenance cost	Minimise
Cost recovery	Financing mechanisms to cost-recover initial capital, and operation and maintenance costs	Maximise

Source: Bhattarai and Neupane (2001).

Planning and assessment of improvement options   77

Table 5.2 Ranking of drainage options according to importance expressed by local stakeholder groups.

	Socio-economic	Technical	Environmental	Managerial	Financial	Aggregate score
Municipality	B, C, A	A, C, B	B, C, A	A, C, B	A, C, B	A, B, C
Public	B, A, C	C, A, B	B, C, A	A, C, B	A, B, C	A, C, B
Slum dwellers	A, C, B	A, C, B	C, B, A	A, C, B	A, C, B	A, C, B
Business	B, C, A	B, A, C	B, C, A	A, C, B	A, C, B	B, C, A
Farmers	B, C, A	A, C, B	B, C, A	A, B, C	A, C, B	B, C, A
Aggregate	–	–	–	–	–	B, A, C

Source: Bhattarai and Neupane (2001).

To assist the decision-making process a computer programme was used to assist the processing of subjective judgements combined with objective values in a single framework about the relative merits of different options. The results of this Analytic Hierarchy Process based on interviews and participatory approaches are presented in Table 5.2.

The analysis showed that a different drainage option was chosen when the interests and concerns of all the stakeholder groups were taken into consideration in comparison to the one that was chosen by the municipality alone. In addition, different stakeholder groups prioritised different options: the business and farming community selected option B, whereas the public and slum dwellers selected option A, the same option as the municipality. Finally, the experiences from the planning process indicated that a stakeholder group can have a distinct set of preferences, but intra- and inter-stakeholder group discussions can enable a negotiated outcome that is envisaged to be better from the perspective of society as a whole.

5.5 INFORMATION COLLECTION AND MANAGEMENT

Information management is critical for decision-making and this requires a good information base in order to make effective decisions during all stages of the planning process. Access to good data will be important as a baseline, which may subsequently be used to monitor the progress of plans and activities.

During the planning process, especially for larger schemes, large amounts of information on the area of development will be collected. Once the data has been collected and the initial plans are drawn up, numerous alterations and updates may be necessary during the project lifetime. It may be that further information is required, which was not collected initially, or the original planned line of a drain becomes unsuitable and has to be amended.

To enable these data to be used effectively to help guide the decision-making and later the design of the urban drainage system, the data needs to be collected and handled in a systematic manner. Therefore, collection, compiling, presentation and storage of this information require careful management to ensure it is up-to-date. The main procedures involve in information management systems involve:

(1) collection, acquisition and verification;
(2) collation and storage;
(3) processing, analysis and comparison;
(4) revision and updating;
(5) retrieval and presentation.

5.5.1 Geographical information systems

Geographical information systems (GIS) are versatile information management tools, which have widespread application in urban catchments, both to assist the development of solutions to runoff problems and for the development of urban planning control measures. GIS may be used to compile spatial data from existing maps, field surveys or remote sensing techniques (satellite imagery and aerial photography). This catchment data can then be combined with other data, such as housing density, water consumption, sanitation facilities, etc. as well as demographic and socio-economic data.

GIS provide a convenient means of storing and visualising this spatially distributed data and may be applied for the planning and design of urban drainage systems as well as other forms of infrastructure. An increasing number of urban upgrading projects have utilised GIS for planning and design of the infrastructure for informal settlements, using information about land use and ownership as well as low-lying areas which are prone to flooding (e.g. Imparato et al. 2000).

In particular, GIS can assist the drainage engineer by storing, processing and presenting spatial information that is relevant to the scheme from its initial planning to full operation. GIS are appropriate for urban stormwater planning due to the spatial and physical natural of urban runoff problems. A map in GIS format with attributes of the area may be used to calculate the impervious areas whilst maintaining an accurate record of geo-referenced data and can be then used to calculate runoff and analyse existing and proposed drainage systems.

However, it is important to recognise the potential limitations of information management systems and it should not be assumed that GIS technologies will offer the answers to urban drainage problems. As with other forms of information technologies, there is a danger that the too much emphasis is placed on the development of a GIS without sufficient consideration on how it will be used and who will use it in the long-term. Lessons from the application of a GIS system in Mirzapur in India suggest that the most important enabling factors relate to the step-by-step introduction of new technologies, training of local staff, the definition of tangible goals of the GIS application and the presence of a person who is suitably skilled to set up the GIS and to ensure that people understand what it is for and how it may be used (Saladin et al. 2002).

5.6 SPATIAL MAPPING AND PHYSICAL INFORMATION REQUIREMENTS

In developing countries, spatial data required for GIS, or even for more simple maps, are often not easy to find and up-to-date and accurate information is generally lacking. This is especially the case for informal settlements where the development has not been planned and there are no records or maps describing the construction of buildings or the infrastructure. A good information base describing land use and topography is vital for the calculation of runoff to enable a good evaluation of the existing drainage system performance and for the design of new systems.

The collection of spatial information will therefore be necessary to construct base maps, which are essential in the planning process. The project area, in relation to natural drainage basins, needs to be established during the initial stages of planning for an intervention to the drainage system. In some cities, defining catchment boundaries is a straightforward process once the topographical maps are produced, but in other cities, characterised by flat topographies, this procedure can be problematic. An entire settlement might conform to a single drainage basin, but where this is not the case the project area may be divided into different basins. Errors in defining catchment boundaries can lead to large-scale errors in the estimation of runoff and subsequently in the design of drainage systems.

It is therefore important to undertake a preliminary assessment of the drainage areas and then to revise these according to results from more detailed fieldwork. These investigations may involve surveys during storm conditions, as this might be the only reliable way of acquiring a good understanding of the catchment boundaries, surface runoff flow pathways, and how the drainage system is performing during flood conditions (Kolsky 1998).

One of the main considerations when choosing a mapping technique should be the scale for the map. The topographical data should be defined at a scale, which is suitable for design purposes or for the development of a computer model. The scale of mapping and the level of accuracy for the mapping will depend on the size of the catchment and, to a certain extent, the topography of the catchment. In general, flat catchments will need a higher degree of accuracy in topographical data than catchments which have greater land slopes due to the fact that errors in elevation can have a greater effect on the spatial distribution of flooding.

5.6.1 Acquisition and collection of spatial data

A combination of survey techniques will be required to develop a comprehensive view of the drainage area in the form of a map. There are a number of alternative approaches towards the collection of data which are described below.

Land-based topographical surveys

Conventional surveying methods such as traversing (to mark the location) and levelling (to measure elevations) are the most common form of mapping. There is now widespread use of 'total station'[1] survey equipment to produce conventional ground-based surveys much quicker than was possible in the past. More advanced surveying methods include the use of geographical positioning systems (GPS), which use satellites to locate ground positions and heights.

On the catchment surface, it will be necessary to obtain elevation and locations at road junctions, roads and roadsides (noting whether the adjacent land is above or below the ground) and general topography of open land (low and high spots). In areas where flooding is experienced or expected, it is important to gather data on the levels at which water can enter houses together with information about water levels during past flood events.

Land-based surveying will be compulsory in order to map the routes of drainage systems and the details of the cross-sectional areas, drain conduit slopes and other information regarding storage basins, etc. In addition, a detailed survey is needed to evaluate the structural condition of the drainage system and the level of sedimentation and blockages caused by solid wastes in the drains.

Aerial remote sensing

Remote sensing (including different types of aerial-based photography and satellite imagery) provides a useful tool as an alternative to traditional ground surveys to collect the data and plans drafted by hand which are labour intensive, time consuming and difficult to organise in some locations (Hunter 2001). Table 5.3 summarises some of the advantages and disadvantages of the available remote sensing sources of data for mapping. Aerial surveying can provide details for assessment of contributory catchment area and land use, which can then be used to calculate runoff. It can also identify some of the large open drainage channels in the catchment, but cannot provide the detailed data that is required to describe the whole of the drainage network.

Remote sensing images are a particularly good way of estimating the level of development and urbanisation to calculate impermeability (as described in Chapter 1, in particular, see Figure 1.2). These are of specific importance in cities in developing countries where the availability of good maps and data is so poor, that the production of base maps is the very first stage in the preparation of urban drainage plans. However, in general, it will still be necessary to use ground surveying techniques to provide the level of detail required for design purposes.

Traditionally, satellite imagery has been used for small scales between 1:25,000 and 1:250,000 using existing systems such as Landsat and SPOT. These

---

[1] A total station is a type of surveying equipment which combines a theodolite with a level and all information is automatically fed into a computer.

Table 5.3 Comparison of remote sensing data acquisition methods.

Source	Advantage	Disadvantages
Aerial photography	High resolution/accuracy Well-established technology Simplicity of processing	High costs and delays Security restrictions Need for a lot of logistics Need for ground control[a]
Small-format aerial photographs	High resolution Low complexity	Not well known Need for ground control[a] Limited to small surveys
Airborne radar (includes LIDAR)	No cloud problems Fast coverage	Low resolution[b] High costs and complexity
Satellite imagery: high resolution (Landsat, SPOT)	Fast production Low ground control/logistics Time series possible	Low to medium resolution Small details not visible Not well known
Satellite imagery: very high resolution (QuickBird, Ikonos and Orbview3)	Fast production Small details visible Low ground control/logistics Good geo-reference Time series possible	More expensive than other satellite images

After Hunter (2001).
[a] Ground control refers to measurements needed for geometrical conditions. It does not mean reference data for the analysis.
[b] LIDAR now commercially available and much greater resolution possible.

systems do not provide the same level of detailed information available with aerial photography and ground surveys at large scales. However, in recent years a number of very high-resolution satellite systems have become operational, which may serve as an alternative to aerial photography. At the time of writing this book, three commercial satellite systems, QuickBird, Ikonos and Orbview 3, are able to provide panchromatic imagery data with a spatial resolution of 1 m or better. Examples of suggested maximum appropriate mapping scales based on the satellite systems mentioned above are listed in Table 5.4 (*Note*: These vary considerably from 1: 40,000 to 1: 1,500).

In addition to the panchromatic imagery data, the sensors also provide multi-spectral data allowing the creation of colour composites. Compared to black and white (B/W) images, this facilitates an easier understanding and interpretation of the images as they resemble traditional colour photographs. Furthermore, the colours in the digital colour images can be manipulated to provide the maximum contrast between different surface types making it easier to identify specific surface types. Instead of the 8-bit greyscale in B/W images with only 256 different greyscales, the colour images can display 24-bit values allowing the display of more than 16 million different colours. Pan-sharpened 'true-colour' images can be created by merging the two data sets described above to create colour images in a 0.6 m spatial resolution (see Figure 5.3).

Techniques exist that enable the transfer of information from panchromatic data to enhance the appearance of the multi-spectral data by increasing the spatial

Table 5.4 Type of satellite imagery, resolution and maximum map scale.

Type of satellite imagery	Resolution	Maximum map scale
SPOT 5	2.5 m B&W pixels	1:5000
	5 m B&W pixels	1:10,000
	10 m colour pixels	1:15,000
Landsat ETM+	30 m colour pixels	1:40,000
Ikonos	1 m B&W pixels	1:3000
	4 m colour pixels	1:8000
QuickBird	0.6 m B&W pixels	1:1500
	2.4 m colour pixels	1:5000

Figure 5.3 Satellite picture of Port Au Prince, Haiti with a resolution of 0.6 m. (*Note*: Original images are in colour as described above.)

resolution. This data product, known as pan-sharpened data, is available from the data providers at extra cost. QuickBird data provides the highest resolutions, featuring four multi-spectral bands at 2.4 m and a panchromatic band at 0.6 m resolution, whereas both Ikonos and Orbview 3 provide 1 m panchromatic and 4 m multi-spectral data.

Another advantage with satellite images for mapping purposes is the development of image archives. Unlike aerial photography, which is limited to the date when the survey was carried out. The image archives also make it possible to perform historical analyses going back in time, which is particularly useful to be able to observe the growth of urban areas. The very high-resolution satellites have a high repeat cycle of only a few days. This means they pass a certain location on the Earth's surface within a few days interval, which makes it possible to collect and process data within a few days. This may be particularly useful in emergency situations where flooding is severe and prolonged.

Furthermore, the ground-processing segment has been drastically improved in recent years. The geo-reference of the very high-resolution images are usually within some metres and the final adjustment of the geo-reference can be done using only a few ground control points. With the introduction of very high-resolution systems (1 m resolution or less), the level of detail that can be achieved has been brought much closer to the more traditional methods used for large-scale surveying and it is now possible to provide the same level of detailed information from satellite data as is available with aerial photography and ground surveys at larger scales (Hunter 2001).

5.7 REFERENCES

Abrishamchi, A. and Tajrishi, M. (2002) Urban water management planning using a multi-criteria decision-making approach, case study: city of Zahidan in Iran. *Proceedings of the International Conference on Urban Hydrology for the 21st Century*, Kuala Lumpur, 14–16 October.

Bhattarai, S. and Neupane, B. (2001) Informed decision making for drainage management. *Water, Sanitation and Hygiene: Challenges of the Millennium, Proceeding of 26th WEDC Conference*, Dhaka, Bangladesh.

Fletcher, T. D., Wong, T. H. F., Duncan, H. P., Coleman, J. R. and Jenkins, G. A. (2001) *Managing Impacts of Urbanisation on Receiving Waters: A Decision Support Framework.* Cooperative Research Centres (CRC), Catchment Hydrology, Department of Civil Engineering, Monash University, Australia.

FWR (1998) *Urban Pollution Management (UPM) Manual – A Planning Guide for the Management of Urban Wastewater Discharges during Wet Weather*, 2nd edn. Foundation for Water Research Report FR/CL 0009, November.

Hunter, J. N. (2001) *An Investigation into Base Mapping for Low-cost Sewerage Using GIS and Alternative Methods of Spatial Data Acquisition.* MSc in Sustainable Management of the Water Environment, Department of Civil Engineering. University of Newcastle upon Tyne, United Kingdom.

Imparato, I., Mingucci, P., Muzzarelli, A. and Petrella, L. (2000) *Putting the Urban Poor on the Map – An Informal Settlement Upgrading Methodology Supported by Information Technology.* UNCHS (Habitat), Nairobi, Kenya.

Kolsky, P. J. (1998) *Storm Drainage: An Engineering Guide to the Low-cost Evaluation of System Performance.* Intermediate Technology Publications, London.

Saladin, M., Butler, D. and Parkinson, J. (2002) Applications of Geographic Information Systems (GIS) for municipal planning and management in India. *Journal of Environment and Development* 11(4), 430–440.

Tayler, K., Parkinson, J. and Colin, J. (2003) *Urban Sanitation: A Guide to Strategic Planning.* Intermediate Technology Publications, September.

Williams, E., Dempsey, P., Crabtree, B. and Walwyn, R. (2003) Simplified integrated modelling of a large conurbation – the River Tame catchment. *Proceedings of the International Conference on Application of Integrated Modelling*, Tilburg, The Netherlands, 23–25 April.

6
Configurations of urban drainage systems

Drainage systems are important components of urban infrastructure, providing an essential service for urban communities living in close proximity to one another. Their main function is to collect and transport stormwater away from human settlements in order to protect health, quality of life and enable daily urban activities to continue (see Figure 6.1). In addition to drainage of stormwater runoff, they may also collect wastewaters derived from various domestic, commercial and industrial activities.

This chapter focuses on various factors that influence the choice of urban drainage system that may be employed to drain runoff from urban areas in order to reduce

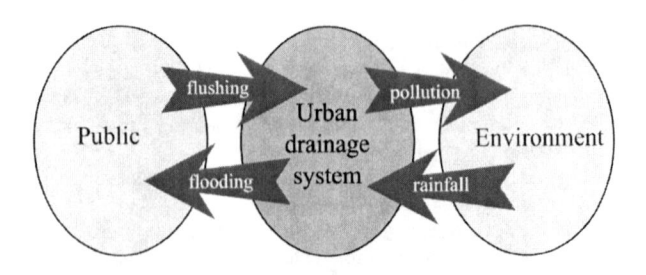

Figure 6.1 Interfaces between urban drainage systems, the public and the environment (Butler and Davies 2004).

© 2005 IWA Publishing. *Urban Stormwater Management in Developing Countries* by Jonathan Parkinson and Ole Mark. ISBN: 1843390574. Published by IWA Publishing, London, UK.

flooding. The chapter refers to quality aspects, but focuses mainly on hydrological and hydraulic aspects. It is intended as an introduction to the variety of technologies that are available, but does not provide detailed design information, which is found in various engineering texts and design guides (see Appendix A1).

6.1 MAJOR AND MINOR DRAINAGE SYSTEMS

As shown in Figure 6.2, one of the main considerations for the drainage of surface runoff influencing the type of drainage system is the capacity of the *minor* drainage system (which consists of constructed pipes and open channels) in comparison with the *major* drainage system (which consists of natural channels and surface flow pathways).

Conventional design procedures for urban drainage systems focus on the minor system, which is the more easily recognisable part of the system consisting of constructed drains and channels, and other forms of flood protection. However, this does not take into account the fact that in tropical climates, where runoff is frequently greater than the capacity of the drainage infrastructure, a significant proportion of runoff occurs as overland flow. The major system drains floodwaters associated with larger rain events and is important because severe damage and personal injury may occur during intense storms if no provision is made for these flows.

Figure 6.3 shows the drainage paths from three catchments during normal flow conditions, but once these are exceeded, the major drainage system predominates and the flow path is dictated, not by the paths of the minor drainage system, but by the topography and land surface conditions. This is particularly apparent in flat catchments where topographical differences are relatively small.

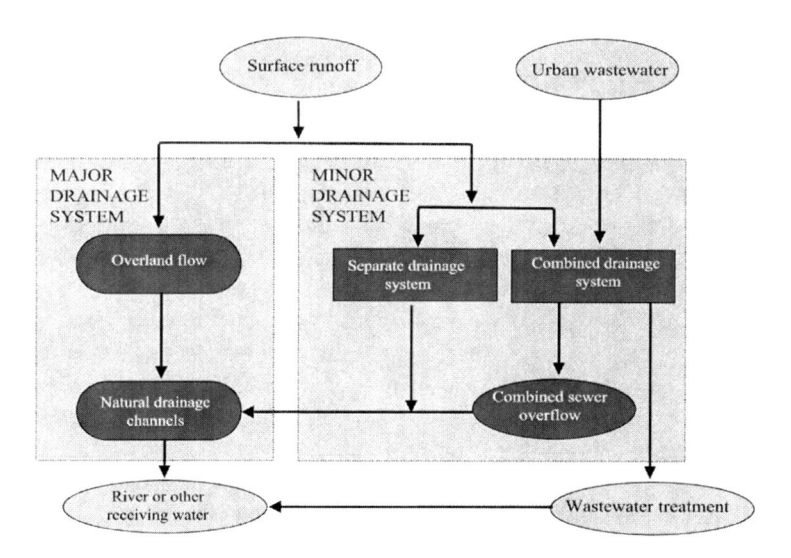

Figure 6.2 Major and minor drainage systems and their interactions.

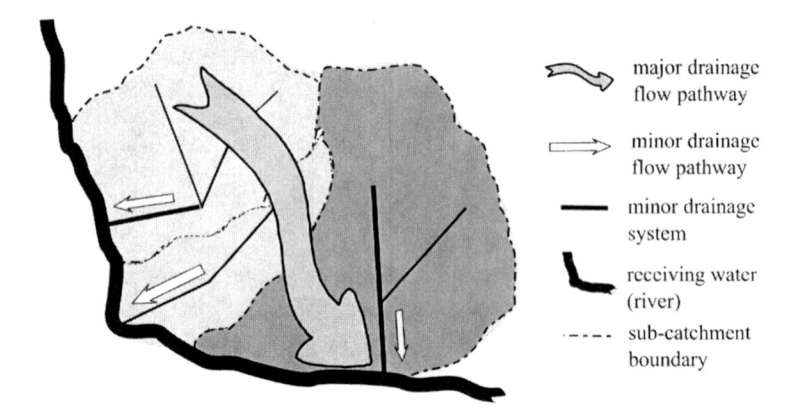

Figure 6.3 Minor and major flow paths in a flat catchment.

The role of the major drainage system is therefore very important and the capacity of roads, carriageways and open spaces form important parts of the drainage system for the control of large-scale flooding. In fact, during heavy storm events, the majority of the runoff may be routed via streets and other overland flow pathways.

The interactions between the minor and major drainage systems during flood conditions are particularly important in tropical and subtropical climates and it is good practice to define the flow paths that operate during storms of different severity. However, most urban drainage designs do not take this into consideration despite the critically important role that these have for drainage of runoff and flood alleviation in the urban environment. It is therefore the responsibility of the drainage engineer to consider both the minor and major drainage systems.

6.1.1 Roads as drains

A novel approach towards stormwater drainage and flood control, which utilises the major flood-routing system, involves the use of roads as part of the drainage system. Although this approach does not negate the need for a minor drainage system, the use of the road carriageway and the road layout itself to convey large flows of stormwater runoff is particularly appropriate in the urban environment. This approach requires less maintenance than conventional drains as street sweeping is easier than drain cleaning and local residents also have an interest in keeping roads clear to enable vehicles to pass and to improve the quality of the local environment.

However, the design of roads to operate as drainage conduits conflicts with conventional road engineering design, which aims to drain runoff as quickly as possible from the carriageway in order to minimise damage to the road surface. Therefore, if a road is designed for flood routing, greater attention is needed to ensure that the road surface is of good quality and requires more extensive site grading in order to provide protection from erosion. Furthermore, if a street is used as an active part of the drainage system, it should be ensured that the depth of the channel in the roadway is not too high (maximum 30–40 cm) because if the flow is too deep it will be dangerous to local residents and also make the road impassable for emergency traffic.

Even though the use of road networks as drainage conduits to convey heavy flood flows is logical for cities in developing countries, which suffer from frequent flooding during the rainy season, there are few examples in the literature describing such a use. One example is in Indore (Madhya Pradesh, India) where a slum improvement project pioneered the use of roads as drains in a concept entitled *slum networking*, which is described in more detail in Chapter 3. To prevent water accumulation in low-lying areas, the roads were constructed to be lower than the surrounding area and the soil from the excavation was used to fill up low-lying areas and reduce the ponding of surface water in these areas.

6.2 SEPARATE AND COMBINED DRAINAGE SYSTEMS

The other important design consideration highlighted by Figure 6.1 is whether runoff is routed through a separate stormwater drainage system or combined with the system for collection, drainage and disposal of other wastewaters from domestic, commercial and industrial sources.

6.2.1 Separate drainage systems

Separate systems consist of two drainage networks, one for surface runoff (wet weather flows) and one for dry weather flows of urban wastewater. The surface runoff is drained by a network of storm sewers and discharged into the receiving waters (usually without any form of treatment). Urban wastewaters are drained separately by a network of sanitary sewers to a treatment plant, which treats the wastewater prior to discharge.

However, the reality is quite different in many developing countries. Separate systems are more expensive to construct; hence a stormwater system is often constructed without provision for the drainage of wastewaters. As a result, the stormwater drainage system becomes a recipient for wastewater and effluents from septic tanks as well as stormwater.

Even in high-income countries, where there are more advanced systems for the regulation of wastewater discharges, there are often many cross-connections and unregulated discharges into the drainage systems. As a result, in the majority of situations in developing countries, where there is little planning control and regulation of the construction of drains, separate systems rarely remain fully separated. The problems of both cross-connections and illegal connections to storm drains result in serious problems due to discharge of untreated wastewater into receiving waters. In addition, discharge of stormwater into systems designed to convey dry weather flow can cause problems for the operational performance of wastewater treatment processes that are not designed to cope with large flows of dilute storm runoff.

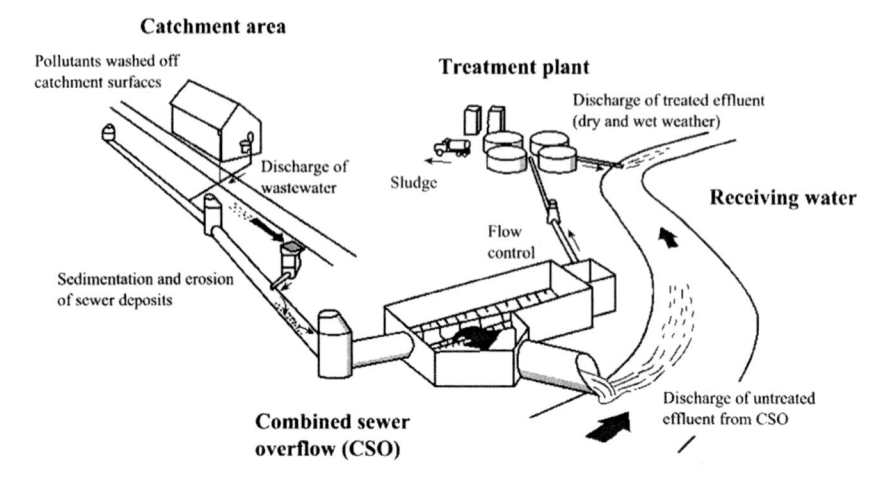

Figure 6.4 Components of the combined drainage system with a high-sided weir overflow (after Weiss 1994).

6.2.2 Combined drainage systems

Combined systems drain a mixture of domestic, commercial and industrial wastewaters during dry weather, and during wet weather they operate as flood control drainage systems. Figure 6.4 illustrates the combined drainage system, showing the main components and processes related to runoff and pollutants. One of the most important features of the combined drainage systems is the combined sewer overflow (CSO). There are various types of CSO, such as hydrodynamic vortex separator, stilling pond and weir, and screens. The relative merits of these different devices are discussed in detail by Butler and Davis (2004).

During storm events, as the rainfall-generated inflow into the combined sewerage system increases, the hydraulic capacity of the collection system is exceeded and subsequently the excess flow is discharged from the CSO into the receiving waters (a river, lake or coastal area). Stormwater runoff is relatively unpolluted, but becomes highly polluting when mixed with other wastewaters or when the high velocities of storm flows erode the sediments deposited in the drainage system. Therefore, as described in Chapter 1, the first flush of runoff can cause a serious deterioration of the water quality in the receiving waters.

6.3 UNDERGROUND AND SURFACE DRAINAGE SYSTEMS

There are basically two types of minor stormwater drainage systems, 'open' channels and 'closed' pipes. However, many drains that are designed and constructed as open drains later become covered, whereas covered drains often become partially

Figure 6.5 Well-maintained culverts in an open storm channel in Ruma, Serbia. (Photo: Slobodan Djordjević.)

open as covers are broken or removed. Thus, in reality systems are often a combination of both; and even in one street, some sections of drain will be covered, while other sections will remain open. In urban areas, open drains may comprise a number of culverts under individual car driveways (Figure 6.5), which may reduce the hydraulic capacity of the drain, but at the same time contribute towards the retention function of the drainage system.

6.3.1 Underground drainage systems

In higher-income countries in temperate climates, drainage systems are often constructed below ground in closed pipes, because they are preferable from a health and aesthetic perspective. In addition to the need to maintain self-cleansing velocities, buried pipe systems require:

- Deeper excavation.
- Structural strength to withstand heavy loads on the street overhead.
- Expensive additional ancillary structures (e.g. manholes and overflows).
- More sophisticated engineering techniques.

From a perspective of operation and maintenance, deterioration and accumulation of debris or sediment are more difficult to monitor/control in closed drains compared to open drains and closed drains are more expensive to maintain. Mosquito breeding in closed drains is more difficult to control and slowly moving sewage may produce hydrogen sulphide, which attacks cement and concrete if it is not well ventilated.

In addition to these construction and operational and maintenance constraints, the high cost of piped systems will usually be prohibitively expensive for lower-income communities in developing countries, and therefore in the majority of situations a covered open channel is the preferred option.

Control of solids entering closed drainage systems

One of the most important design features of closed drainage systems is the type of inlet, which allows for the inflow of runoff but avoids the ingress of solids into the system. For covered/closed systems to be efficient at draining stormwater runoff, they need to have inlets, which are:

(1) large and numerous enough to allow for the inflow,
(2) not so big that they become dangerous,
(3) effective at stopping the ingress of solid waste.

These may include a wide range of devices such as grids or gully pots. However, if they are effective at keeping solids from entering the drainage system, the inlets invariably become blocked. Therefore, it is of vital importance that there is a routine maintenance system to ensure that they are regularly cleaned – especially before and during the rainy season (Chapter 9).

Solid waste traps may be installed at strategic locations in stormwater drainage systems to collect and remove solid waste from the flow. There are many different types of solid waste traps, and the efficiency at removing solids is highly variable, but the majority of them rely on screening of stormwater runoff, vortex flow to remove solids by centrifugal force, or surface debris traps to remove floating debris.

The main criteria determining the suitability of a particular trap will be the flow rate, allowable head loss, size, efficiency, reliability, ease of maintenance and cost-effectiveness. However, the choice of trapping structure is site-specific and the location of the traps is crucial. Decisions need to be made to include one or two large solids traps towards the end of the drainage system or to install smaller ones at strategic points in the network. The efficiency will rapidly be lost if these traps are not properly cleaned and maintained, and therefore easy access for cleaning and maintenance is crucial. In some instances, the cost of providing adequate access may be higher than the structure itself (Armitage and Rooseboom 2000).

6.3.2 Surface channels

Surface drains are generally easier and cheaper to construct compared to buried pipe systems. This is particularly the case when dealing with large stormwater

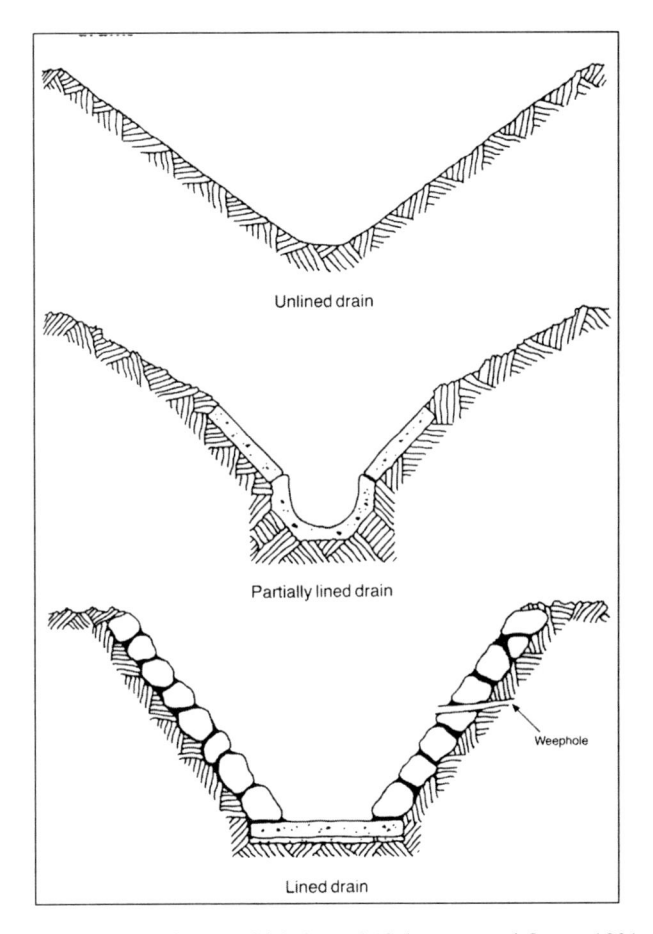

Figure 6.6 Various types of trapezoidal channel (Cairncross and Ouano 1991: reproduced with permission of WHO).

discharges and in cases where land subsidence is a problem. Figure 6.6 shows various types of cross-sections of different open channels. If the surface drains are open, they often act as recipients of solid waste dumped by local residents, which may reduce the flow capacity. In addition:

(1) They are not as hygienic as closed drains and may smell.
(2) Children may fall into them or play next to (or in) them.
(3) They may be breeding grounds for mosquitoes.

However, open drains are less prone to blockages than pipes, and are easier to inspect and access for removal of debris. Furthermore, when drainage problems occur, it is easier to identify and rectify the location of the problem in an open

Figure 6.7 A trapezoidal stormwater channel in Hanoi, Vietnam, receiving wastewater effluents from septic tank overflows. (Photo: Jonathan Parkinson.)

drainage system. It is also easier to locate connections from wastewater discharges such as the one shown in Figure 6.7 as part of a pollution control strategy.

6.3.2.1 Drainage channels for steep slopes

Drains for steep slopes need special consideration as high intensity of rainfall causes high flows with high velocities, which can result in a high erosive capacity. Therefore, as shown in Figures 6.8 and 6.9, drains need to be constructed with some form of energy dissipation. Drainage systems for high-density residential areas can be combined with local streets and paths in the area. An example can be seen in Figure 6.10 showing a drainage system in São Paulo in Brazil, which combines the drainage channel with the pedestrian pathway and includes an innovative design for a self-cleansing solid waste trap.

(a) Baffles

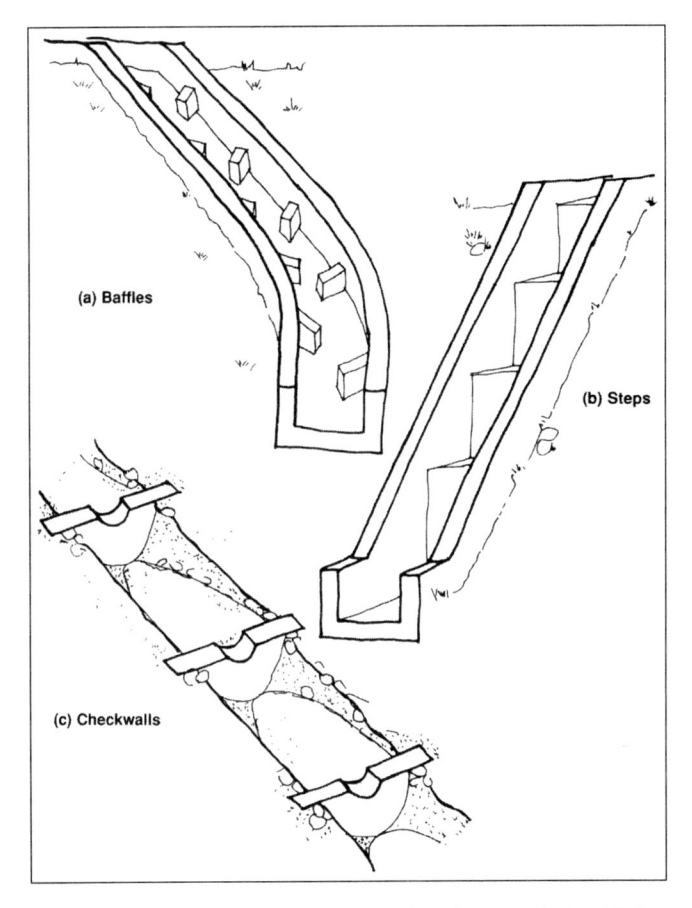

(b) Steps

(c) Checkwalls

Figure 6.8 Various types of energy dissipation devices for steep drains (Cairncross and Ouano 1991: reproduced with permission of WHO).

6.4 ATTENUATION OF STORMWATER RUNOFF

The storage function of drainage conduits is generally not considered to be significant enough to be included in most designs. However, it can offer a considerable amount in terms of flood alleviation, and the complex hydraulic relationship between storage and flow is critical to the overall function of the drainage system.

In addition to the storage in the system itself, stormwater retention and detention ponds (otherwise known as basins) are an effective way to control stormwater runoff and are widely used as flood protection devices. Figure 6.11 shows the most important impacts of the runoff hydrograph associated with the introduction of attenuation ponds. The reduction in peak runoff discharge and increased lag time (time after rainfall to reach peak runoff) has a number of beneficial effects related to a reduction in flooding and discharge of pollutants into receiving waters.

Figure 6.9 Steep drain in São Paulo, Brazil. (Photo: Jonathan Parkinson.)

As shown in Figure 6.12, there are two main types of detention and retention ponds (on-line and off-line) and these can be open or covered. Both types offer the same storage function, attenuate peak discharges, and contribute towards the control of flooding and discharges from overflows.

6.4.1 Detention basins

Detention basins are storage tanks or areas that are designed to 'detain' runoff and then drain completely after stormwater runoff ends and become dry between storms. Detention basins may be lined or unlined, with the unlined ponds usually being vegetated. The extended detention basin uses a much smaller outlet, which extends the retention time for more frequent events so that pollutant removal is facilitated via sedimentation.

Configurations of urban drainage systems 95

Figure 6.10 Drainage design in the Parque Amélia Favela; one of the 32 favelas that benefited from the Guarapiranga Project, São Paulo, Brazil. (Photo: Heitor Cesar Riogi Haga.)

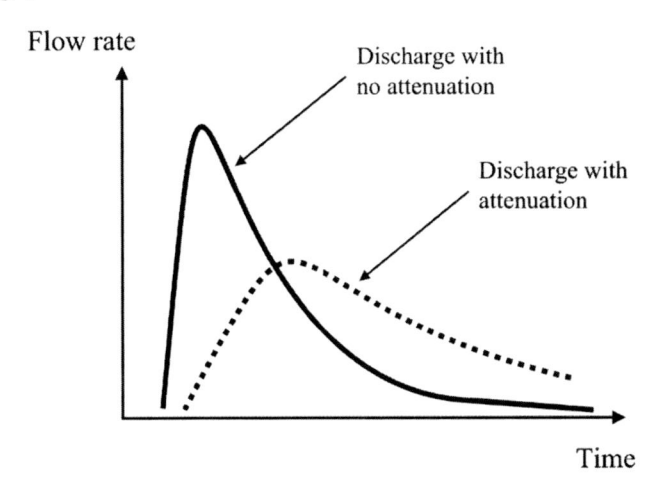

Figure 6.11 Attenuation of runoff in a storage pond (typically ponds release the water over a period of 12–36 hours).

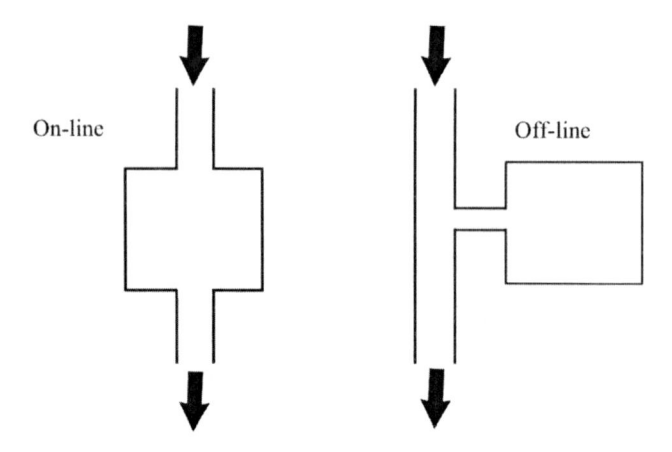

Figure 6.12 Schematic diagrams of 'on-line' and 'off-line' stormwater attenuation ponds.

Stormwater detention systems must operate over a wide range of hydrodynamic and pollutant loading conditions. The design typically includes the specification of a uniform detention time for extended detention ponds to ensure water quality control. However, in practice, stormwater pollutant detention periods vary according to the system hydrology (e.g. inflow characteristics and antecedent storage conditions) and the characteristics of the pollutographs. Hence, performance of detention systems for the removal of suspended solids and associated pollutants from stormwater is highly variable.

Control of the development is also important for areas designated for detention of stormwater runoff. Innovative approaches to land control, which have positive benefits during flood conditions, may be employed. For example in Brazil, as shown in Figure 6.13, areas designated for flood control are also used for sports and amenities to discourage illegal invasions and squatter settlements (Tucci 2002).

6.4.2 Retention basins

Retention basins are sometimes referred to as *wet ponds* because they retain water permanently. A dry weather base flow from a natural drainage system (e.g. a small stream), a pond liner and/or high groundwater table is generally required to maintain the permanent pool. The water in the pond is displaced and replaced in part or in total by stormwater during a storm event. The other losses are as a result of infiltration into the soil and due to evaporation.

Figure 6.14 shows some examples of large retention basins in Bangkok, Thailand. In addition to the hydraulic performance of the basin (flow and flood control by reduction of flow peaks), the water quality will improve due to sedimentation of sand, silt and clay and, in particular, the capturing of the first flush of

Figure 6.13 Detention pond designed as a football pitch, Porto Alegre, Brazil. (Photo: Carlos Tucci.)

Area 28.8 Hectare Area 43.52 Hectare Area 13.76 Hectare
Storage 1.8 M m³ Storage 3.5 M m³ Storage 0.6 M m³

Figure 6.14 Retention basins in Bangkok, Thailand.

sediments and associated pollutants. However, the variability of the reduction of pollutants will depend on the dimensions of the basin and location of the inlet and the outlet.

In addition to their function as stormwater runoff control devices, stormwater ponds may also serve for aesthetic and recreational purposes. Ponds may also be combined with infiltration and stormwater reuse as described in Chapter 7. The design of retention facilities must clearly address issues associated with how the pond is integrated into the urban environment and how the public perceives it in relation to safety for the people in contact with the ponds and aesthetics of the pond and its surroundings. Maintenance of ponds is therefore very important, especially to reduce the ingress of sediments and solid waste (see Chapter 9).

6.5 URBAN HYDROLOGY DESIGN CONSIDERATIONS

This section briefly discusses rainfall data requirements and the estimation of runoff for the design of urban drainage systems. The flow rate of runoff is the main parameter, which influences the design of urban drainage systems, but pollutant concentrations must be taken into account in relation to disposal of wastewater and stormwater.

The return frequency of flooding is the most important design parameter, which is used to calculate the dimensions of the drainage system. As shown in Figure 6.15, this is the first stage in the urban drainage design procedure, which must take into account local considerations related to economics, public health and pollution control as well as climatic factors which affect rainfall distribution (as described in Chapter 1).

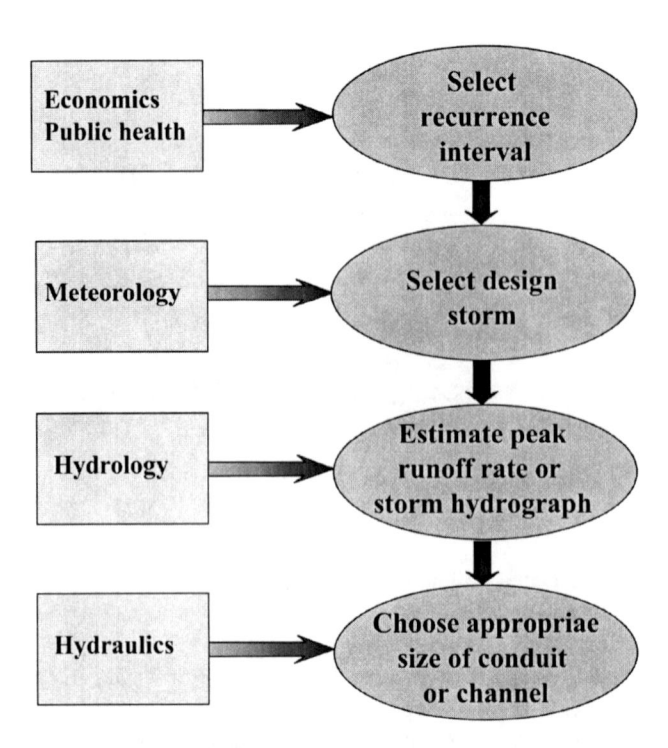

Figure 6.15 Design procedure for the sizing of drainage conduits (Hall 1984).

6.5.1 Rainfall data

The primary factor that governs the volume of runoff is rainfall – notably intensity and duration. Rainfall intensity is often the cause of flooding and is critical for small urban catchments with a short time of concentration. Due to the high rainfall intensities in tropical and subtropical climates associated with convective rain (see Chapter 1), runoff from urban catchments is 'flashy'. The temporal and spatial variations in rainfall intensity can be significant and thus make it difficult for urban drainage engineers to obtain detailed and accurate rainfall data as design parameters. In general, in developed as well as developing countries, the sort of detailed meteorology that is necessary to account for these variations is usually not available without detailed rainfall surveys.

Therefore, it is necessary to base designs on the best available historical records of rainfall data collected from the nearest meteorological station. Based on these data, storm events can be ranked according to duration and intensity, and Intensity–Duration–Frequency (IDF) curves can be derived to provide a graphical representation of the rainfall distribution for a specific location, which can then be subsequently used for drainage designs.

As a general rule-of-thumb, it is necessary to have measured rainfall for a period of at least three times longer than the return period for the analyses; that is, if you are interested in a rainfall event, which has a return period of 5 years then you must have 15 years of rainfall data. However, in developing countries, local rainfall data are rarely available for such long periods. Hence, in order to be able to carry out the statistical analysis, it is necessary to derive standard IDF.

An alternative approach is to derive a synthetic design storm. The wisest thing is to leave this to the local meteorological department. Note that precautions must be taken and a sensitivity analyses carried out if no historical rainfall data are available and a design storm has not already been derived and successfully applied in the past.

6.5.1.1 Runoff and overland flow

A prediction of potential flooding and its consequences requires estimation of a design hydrograph. The Rational Method is the simplest approach used to estimate runoff based on knowledge of rainfall and catchment surface characteristics. This approach may be applied to design urban drainage systems for very small urban developments, but in larger areas the relationship between rainfall and runoff is more complex and cannot be approximated by such a simple relationship. The Rational Method to estimate peak runoff has been criticised as being archaic and it has been recommended that it is replaced by 'data-centred approaches' that use more sophisticated technologies and computer design tools (Heaney *et al.* 2002), such as those described in Chapter 8.

An accurate estimation of the quantity and rate of runoff from urbanised catchments is a complex process and many mathematical models assist in the design of urban drainage systems. The application of computer models enables a more accurate representation of rainfall-runoff processes (surface wetting, surface ponding, infiltration, evaporation and flow routing etc.). A range of rainfall-runoff simulation models is available describing the runoff from impervious and pervious urban areas. These provide the drainage engineer with powerful tools for simulation of urban runoff that can be used for design of new systems or analysis of existing system (see Chapter 8).

However, although data-centred approaches and computer models offer drainage engineers with powerful design tools, they should be applied with caution. Care should be applied before using these models as the use of erroneous values or inaccurate assumption of design parameters in hydrological models in order to calculate the resultant volume of runoff and peak flows can lead to large-scale errors in the design of drainage systems. Inadequate data for computing design parameters, and the use of various assumptions and mathematical formulations more relevant to developed countries can create more complex problems rather than solutions to drainage problems in developing countries (Packman 2000). In particular, the use of default parameters derived from analysis of catchments and drainage systems in the industrialised countries in temperate climates can result in large-scale and cost errors due to the construction of drainage systems which are inappropriate for the local situation. Therefore, a lot of research and development is still necessary to develop practical approaches, local understanding of the dominant physical processes and alternatives to 'Western' formulae and design parameters.

Due to the potential risk of failure or under-design of drainage systems, engineers often deliberately use runoff coefficients higher in cities in the developing world to compensate for lack of data and to avoid the uncertainty of designing a system that is inadequately sized. However, providing increased flow capacity for runoff that in reality may never reach the drainage system is an unnecessary expense and the extra costs may mean that equally vital schemes in other areas have to be cancelled or delayed.

6.5.1.2 *Return frequency of flooding*

Stormwater drainage systems are conventionally designed to drain urban runoff from built-up areas so that flooding should occur no more frequently than once during a specified return period. The return frequency or return period of flooding is the most important parameter used for planning and design of urban drainage systems for flood protection, but there is often ambiguity between this and the return frequency of rainfall events. These parameters are different because the hydrological response of a drainage system is *non-linear*. This means that a rainfall event with a 5-year return frequency does not result in a flood event of 5-year return frequency.

Storms of specific return frequencies are used for design purposes and to analyse the hydraulic performance of a drainage system under different wet weather conditions. In addition to return frequency of flooding, these design storms may be used to assess the relative frequency of surcharge in the drainage system

Table 6.1 Guideline flood return frequencies for design of urban drainage systems.

Land use	Return period (years)
Public spaces	1–2
Suburban residential districts	2–5
Higher-density residential areas	5–10
City centres and high-value districts	10–20
High-risk areas	20–50

or the frequency of operation of overflows. Therefore, when a design is carried out, it is important to state which of these design parameters is used.

The choice of return periods depends on the land use and the potential consequences of flooding. Table 6.1 provides an indication of the ranges of return periods that are commonly adopted for drainage designs. However, these should not be applied without consideration of the local situation and these design parameters should be discussed and agreed during the planning process (see Chapter 5) before proceeding with the design.

6.6   CAPACITY OF DRAINAGE CONDUITS

Although the concept of a drain itself is very simple, drainage networks are complex systems involving processes of hydrology, hydraulics, solids and pollutant transport mechanisms. Therefore, the design of these systems needs considerable expertise that requires an understanding of these processes. Broad generalisations and assumptions in the design process may result in gross errors of judgement and investments in inappropriate infrastructure.

The majority of stormwater systems are drained by gravity, but in situations where this is not possible a pumping station may be the only solution. For example, this may be the case at the end of the rainy season where the water level in the river is higher than the street levels inside the city. The principles behind the design and operation of pumps and pumped systems are generically applicable for urban drainage systems and described elsewhere (e.g. Butler and Davis 2004) and therefore not discussed in this chapter.

Where the stormwater flow is drained by gravity, the cross-sectional area and the hydraulic gradient of the flow surface are the two main parameters for estimating drainage capacity. However, in steady-state flow conditions with no backwater effects, the gradient of the drainage conduit can be used for estimation of drainage capacity. It is important to remember that, under flood conditions, drainage systems are *surcharged* which means that the capacity of the system is insufficient to transport the runoff under normal free-surface conditions. Under these conditions, water surface levels dictate the hydraulic gradient, and it is the hydraulic gradient that dictates the flow in the pipe, not the gradient of the pipe.

The other consideration is the build-up of sediment and solid waste. The concept of *self-cleansing velocity* is commonly used in engineering designs for the

102   Urban stormwater management in developing countries

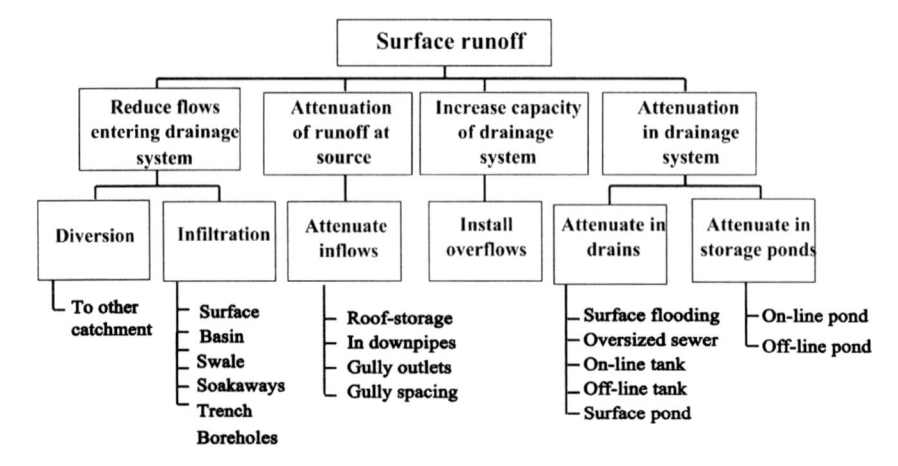

```
                              Surface runoff
        ┌──────────────┬──────────────┬──────────────┐
   Reduce flows    Attenuation    Increase capacity   Attenuation
 entering drainage  of runoff at     of drainage     in drainage
     system          source           system          system
   ┌────┴────┐                    ┌────┴────┐     ┌────┴────┐
Diversion Infiltration Attenuate  Install    Attenuate in  Attenuate in
                       inflows   overflows    drains       storage ponds

To other   — Surface   — Roof-storage   — Surface flooding   — On-line pond
catchment  — Basin     — In downpipes   — Oversized sewer    — Off-line pond
           — Swale     — Gully outlets  — On-line tank
           — Soakaways — Gully spacing  — Off-line tank
           — Trench                     — Surface pond
             Boreholes
```

Figure 6.16 Options for control of runoff. (Adapted from Leonard and Sherriff 1992.)

channels and pipelines. This design parameter is used to reduce sedimentation in the drainage conduits, but in reality very few drains are self-cleansing due to the excessive solids loading and long dry periods in developing countries (see Chapter 8). In addition to the self-cleansing velocity, the erosion criterion is used to protect the drainage system from high velocities causing erosion and damage to the structural integrity of the drainage system.

The complexities of the system make it difficult to make accurate estimates of flow capacity because the hydraulic analysis of sections of pipes in isolation does not allows for estimates of discharge under surcharged conditions or backed-up flow. Therefore, computer simulation models are required to analyse the system and provide the confidence that is needed to design drainage systems. The application of computer models is described in detail in Chapter 8.

These models may be used to simulate a wide variety of potential improvement options for the control of runoff, such as those illustrated in Figure 6.16. The most important aspect related to this figure is the fact that increases in the capacity of the drainage system should be seen as the last option after all the other options have been considered. In particular, attenuation of runoff is an important strategy (see Section 6.5) and other forms of source control (e.g. infiltration, reuse) of runoff are described in the following chapter.

6.7 REFERENCES

Armitage, N. P. and Rooseboom, A. (2000) The removal of urban litter from stormwater conduits and streams. Paper 1 – the quantities involved and catchment litter management options. *Water SA*, **26**(2), 181–187.
Butler, D. and Davis, J. W. (2004) *Urban Drainage*, 2nd edn. Spon Press, London, UK.
Cairncross, S. and Ouano, E. A. R. (1991) *Surface Water Drainage for Low-Income Communities*. WHO/UNEP, World Health Organization, Geneva, Switzerland.

Hall, M. J. (1984) *Urban Hydrology*, Elsevier Applied Science Ltd.
Heaney, J. P., Sample, D. and Wright, L. (2002) *Costs of Urban Stormwater Control.* USEPA Report EPA/600/R-02/021. Urban Watershed Management Branch, Edison, New Jersey, USA.
Leonard, O. J. and Sherriff, J. D. F. (1992) *Scope for Control of Urban Runoff.* Vol 3: Guidelines. Report R124. Construction Industry Research and Information Association (CIRIA), London.
Packman, J. (2000) *Dissemination of Guidelines on Urban Runoff Control in the Developing World.* Centre for Ecology and Hydrology, Wallingford, UK.
Weiss, G. (1994) CSO practice and strategies: the German approach. *IHE Seminar on 'CSO – an European Perspective'*, 24th March, IHE, Delft, Netherlands.

7
Ecological approaches to urban drainage system design

This chapter discusses strategies that promote a more sustainable approach towards the management of urban runoff. These strategies and technologies (sometimes referred to as 'Sustainable Urban Drainage systems' or 'Best Management Practices') utilise a variety of different control measures, which aim to reduce pollution problems, conserve natural water resources and also enhance the amenity value of watercourses in the urban environment. In this chapter the use of the word 'sustainable' refers to environmental sustainability, that is, protection of the environment and preservation of ecological integrity for future generations. The other focus of sustainability, which is the focus of Chapter 9, is the more fundamental requirement of operational sustainability.

7.1 STRATEGIES FOR SUSTAINABLE URBAN DRAINAGE

The objectives of sustainable urban drainage are summarised in Table 7.1 and these are closely linked with the principles of integrated water resource management (IWRM) described in Chapter 3. The design of sustainable urban drainage systems is based on principles of ecological engineering, which aim to preserve natural drainage patterns and emulate the natural hydrological cycle. These objectives are

© 2005 IWA Publishing. *Urban Stormwater Management in Developing Countries* by Jonathan Parkinson and Ole Mark. ISBN: 1843390574. Published by IWA Publishing, London, UK.

Table 7.1 Objectives of sustainable urban drainage.

Objectives	
Reduce runoff and protect urban areas from flooding	Minimise changes to the hydrological characteristics of a catchment caused by new developments and introduce technologies that aim to restore natural runoff flow characteristics.
Conserve water resources	Promote infiltration of rainwater to replenish groundwater, and utilise technologies to collect and store runoff for various low-grade applications.
Protect quality of water resources	Minimise the amount of pollution entering the stormwater system via the implementation of source control and reduce pollutant concentrations using appropriate treatment technologies.
Preserve natural habitat and biodiversity	Maximise the extent of flora and fauna in and surrounding urban watercourses to promote nature conservation and biodiversity.
Promote amenity value of water in the urban environment	Encourage the use of urban watercourses as areas for leisure, amenity and environmental awareness.

inter-related and should be considered as components of an integrated strategy for sustainable stormwater management. However, the relative importance of each objective will vary according to the specific location and the demands of the local stakeholders in the catchment area.

7.1.1 Source control

Prevention or mitigation of runoff problems at source (both in terms of quantity or quality) is regarded as a key principle for the design of sustainable urban drainage. Source control is a key principle of sustainable stormwater management, which aims to reduce stormwater problems at the point of generation through the use of structural and non-structural management strategies.

One of the main advantages of source control is the reduced runoff (both in terms of total runoff and peak runoff), and this can translate into a reduced need for investment in downstream infrastructure. From a perspective of control of runoff this involves the use of infiltration and reuse of stormwater runoff, which results in other benefits related to groundwater recharge and reduced demands for water supply.

Source control is also the most effective strategy for pollution mitigation, which involves measures to reduce the contamination of runoff by various pollutants that are commonly found in stormwater discharges.

7.1.2 On-site and off-site technologies for control of runoff

There are numerous types of technologies that can be employed either on-site or off-site to control runoff and pollution, which form part of an integrated solution to stormwater management rather than applying them in isolation.

Table 7.2 Comparison of the benefits of various technologies.

	Runoff control	Water resources	Pollution control	Amenity	Conservation
Runoff reuse	✓	✓			
Permeable paving and infiltration technologies	✓	✓	(✓)		
Swales	✓	✓	✓	(✓)	
Constructed wetlands	✓	✓	✓	✓	✓

Table 7.2 summarises the benefits of these technologies and Table 7.3 provides an overview of their mode of operation in relation to their location in the catchment. The choice of technologies adopted will be driven by a number of factors including type and size of development, physical and environmental constraints and sensitivity of receiving waters. However, it is generally more beneficial to utilise a range of technologies as part of a comprehensive stormwater management strategy.

Some technologies such as swales and constructed wetlands incorporate the use of vegetation, which improves the quality of stormwater runoff by trapping suspended solids and related pollutants. Table 7.4 summarises the pollutant removal mechanisms that are incorporated into various technologies for control of pollution from stormwater runoff. These encompass a combination of physical, chemical and biological processes which include biodegradation or immobilisation of the pollutants. Sedimentation is the principal mechanism through which pollutants are removed in detention or retention ponds, whereas filtration is the principle mechanism in infiltration systems. Wetlands, on the other hand incorporate a complex combination of sedimentation, filtration as well as biochemical processes. As a result, the extent of removal will vary significantly according to the type of technology and the characteristics of the runoff, and the reader is encouraged to consult other sources of information for further details (see Appendix A1).

Vegetation has an important function in the natural water cycle storing water by interception on leaf surfaces and water uptake by plants. Some groups of plants are resistant to both flooding and drought conditions, and these are logical choices for use in stormwater facilities. Good root structure breaks up soils increasing permeability and allowing water to infiltrate. Plants with large root systems are therefore well-suited as the majority of plant biomass is below the ground, which stabilises soils to prevent erosion, and helps plants survive dry periods as well as contributing to infiltration (Girling *et al.* 2000). The use of vegetation, as well as contributing towards runoff and pollution control, has the added benefit of contributing towards the preservation of natural habitats for wildlife. It therefore offers improved amenity value of stormwater runoff control devices, for instance, where attenuation ponds are located in parks.

Ecological approaches to urban drainage system design

Table 7.3 Various technologies for control of runoff and pollution from urban areas.

Source control	Description
Rainwater reuse	Reuse of rainwater on-site by household collection and storage of roof runoff or off-site collection in ponds for communal low-grade use.
Permeable paving	Permeable paving utilises porous materials for urban surfaces (e.g. porous asphalt, or concrete), which infiltrate stormwater into the surrounding soils.
On-site	
Infiltration trench	Infiltration trenches are ditches filled with porous media designed to encourage infiltration of runoff to groundwater.
Soakaway	A soakaway is filled with a porous media, which may require drilling through impervious layers to reach lower pervious layers, which allows surface water to infiltrate into groundwater.
Swale	Swales are open drainage channels lined with vegetation which increase channel roughness and act as pollutant reduction devices.
Off-site	
Infiltration basin	A type of detention basin that stores stormwater runoff to promote infiltration and restore groundwater resources.
Constructed wetland	Wetlands are areas designed to store and attenuate flow with the use of reeds or other type of wetland vegetation to trap sediments and associated pollutants.

Table 7.4 Pollutant removal processes.

Sedimentation	Sedimentation is the primary pollutant removal mechanism in urban drainage structures where reduced velocities allow for particulates to settle. Although in general, sedimentation is limited to larger particles, extended detention allows smaller particles to agglomerate into larger ones and then settle.
Filtering and adsorption	Pollutant concentrations are reduced as (surface runoff) filters through the soil or through a constructed sand bed. Particulates are removed at the ground surface by filtration, while soluble constituents are adsorbed into the soil as runoff infiltrates into the ground.
Straining	Grasses and other plants strain out particulates when stormwater runoff flows over vegetated areas.
Biological uptake	Plants and microbes require soluble and dissolved constituents such as nutrients and minerals for growth and thus, these are removed from the water flow by bacterial action, phytoplankton growth, and other bio-chemical processes.

7.2 RAINWATER REUSE

As a result of increasing pressures to conserve water resources, there is increasing focus on the reuse of urban stormwater runoff. Otherwise known as rainwater harvesting, water shortages in many cities have forced many people to start collecting and reusing rainwater runoff from roofs. Even in countries that have abundant water resources there are invariably times and places where water shortages occur and rainwater systems can alleviate this water shortage (Gould 2000). Ironically, it may even be during the times of extreme flood conditions when other water supplies are disrupted (see Chapter 10) that rainwater is a particularly valuable source of water.

At the household level, on-site stormwater reuse of roof runoff stored in rainwater tanks may supplement others sources of water (see Figure 7.1). At the communal level, runoff may be collected in larger tanks (as in the example from Bangalore described in Box 7.1) for a wide variety of domestic purposes as well as for irrigation and for livestock (Vishwanath 2001).

Pollution levels in runoff will probably mean that in urban areas rainwater will mostly be used for secondary water uses. It is particularly appropriate as a low-grade

Figure 7.1 Rainwater harvesting in Bangalore – The roof is about 50 m^2, the tank capacity is 500 l and estimated collection of rainwater is 23,000 l per year. (Photo: Rainwater Club, Bangalore, India.)

water source for toilet flushing, garden watering and car washing. However, in some locations, generally poorer households, rainwater may provide a source of drinking water, but care must be taken to ensure that the householders are aware of potential health risks via filtration and preferably that the water is boiled or

> **Box 7.1 Rainwater reuse in Bangalore, India**
>
> Bangalore is the capital of Karnataka state in South India and has a population of approximately 6 million. In the early 1960s the city had more than 250 man-made lakes or rainwater 'tanks' in different parts of the city, which were used traditionally by local residents as a source of water for a wide variety of uses. However, since 1971, the population explosion and rapid urbanisation has resulted in many of these tanks being filled up and converted to other forms of land use. As a result, at the last count, only 85 tanks remained.
>
> Due to the increase in paved area in the city centre, intense rainfall and the absence of water bodies to attenuate and store peak flows of runoff, flooding is now a frequent occurrence in many parts of the city during the monsoon season. Interestingly, many of these flood areas are observed to be where the old tanks were located. Concurrently, there is difficulty in supplying the drinking water requirement to the city since the source of water for the city is a river located 95 km away and the water has to be pumped to a pressure head of 500 m. Due to the difficulties of water supply in the city, a curious paradox has emerged whereby the city may be flooded during heavy rains, but there may be little water available in the water distribution system. As a result of this situation, water in Bangalore is the most expensive in India.
>
> With a view to managing water holistically and to reduce flooding, an integrated approach toward stormwater management based on a strategy of source control and reuse in combination with a programme of cleaning and rehabilitation of the drainage system has been adopted. Through sustained pressure from various civil society and non-governmental organisations (NGOs), the State Government of Bangalore commissioned a systematic survey of the tanks and constituted an independent institution called the Lake Development Authority to oversee the rehabilitation and rejuvenation of all remaining tanks in the city. Already funds have been mobilized and many tanks have been desilted and structurally rehabilitated.
>
> In addition, through an amendment to the existing building bye-laws, new houses and other developments are being encouraged to incorporate rainwater harvesting systems in the construction to collect rooftop rainwater and either store it in tanks or recharge the groundwater aquifer. Older constructions are also being encouraged to adopt rainwater-harvesting systems. It is estimated that up to 20 mm of rainfall can be captured from the rooftops of these new constructions during the beginning of a rainfall event, thus moderating the impact of the storm as well as providing supplementary source of water for local residents.
>
> *Source*: Rainwater Club, Bangalore (2004)

chlorinated. Alternatively, storage tanks may be connected to soakaways, which will replenish groundwater. As well as reducing demand peaks imposed on the water supply distribution network, rainwater reuse can offer benefits in terms of attenuation and reduction of peak flows of stormwater runoff, which may also reduce associated pollutant loads from stormwater runoff.

Dhar Chakrabarti (2001) describes how an integrated system of rainwater reuse in Delhi, India incorporates rainwater reuse at the domestic and neighbourhood level in combination with other rainwater reuse techniques at the city level for recharging the aquifer. The results demonstrated that the adoption of these strategies has helped to raise the groundwater level by 3 m in each successive monsoon. The example in Box 7.1 describes the experiences from Bangalore in which rainwater reuse is being actively encouraged by the state government and, also in India, the example in Box 7.2 describes how runoff from city roads may be used to recharge groundwater after treatment. The lessons from the pilot project also suggest that an appropriate regulatory and incentive mechanism needs to be developed and introduced by the city governments to provide the enabling environment (see Chapter 4) to support the implement of the technology.

7.2.1 Practical and economic feasibility of reuse in urban areas

The reuse potential and the feasibility of a reuse scheme in an urban area is site-specific and affected by a range of factors. The urban layout (e.g. population density, open space and housing type) will influence the amount of runoff and the applicability of collection systems. There may be considerable constraints in urban areas

---

Box 7.2 City roads may harvest rainwater: Delhi, India

Delhi is considering a plan to harvest rainwater from city flyovers and highways to replace unusable and saline groundwater. Harvesting would be done during the monsoon after the first rains have swept the roads clean. Oil from the roads would be filtered out in special desilting chambers. The High Court recently made rainwater harvesting mandatory in Delhi following public interest litigation by the NGO Tapas. Now V. K. Jain, founder of Tapas, says that the Central Ground Water Board believes that water could feasibly be harvested on roads and flyovers. The first monsoon showers would clean the roads. Later showers would be sent through pipes to a desilting chamber with an oil separator. The water would be cleaned and then flow into the earth. Delhi authorities are considering starting the scheme on the link road from National Highway 8 to Dwarka, which has little tap water and saline groundwater. The saline water would be drained off to make room for the rainwater.

Source: Saurabh Sinha, Hindustan Times/Water Observatory
http://www.waterobservatory.org 29 October 2004

particularly as roof sizes can be quite small (as low as 9 m^2) (DTU 2002). Due to climatic variations in rainfall distribution, rainfall-runoff reuse may also be limited (Mitchell *et al.* 1999). An increase in outdoor water use during drier months of the year causes a seasonal rise in water demand. However, in more humid climates it is possible to make do with relatively small tanks if one accepts that not all of the household water demand will always be met (Gould 2000).

As well as physical constraints, the feasibility of stormwater reuse is controlled by economic considerations. Financial issues as well as the quality and reliability of existing water supply will have a significant influence on the economic viability. Rainwater reuse on a modest, domestic scale is a comparatively expensive method of obtaining water of potable quality, with water storage tanks representing the most significant cost of a domestic system. The cost factor is particularly significant in countries such as Bangladesh where the dry period lasts up to 6 months (Wehrle 1999). Nevertheless, rainwater reuse is considered to be an important component of an integrated approach towards the management of urban runoff.

7.3 INFILTRATION OF STORMWATER

Direct infiltration into the ground is an important source control strategy for stormwater management, provided the soil is sufficiently pervious and the quality of the runoff is not considered likely to be a cause of groundwater pollution. Infiltration of relatively uncontaminated runoff from rainfall reduces the volume and rate of runoff and recharges groundwater, which can subsequently help maintain base flows in rivers and provide sources of water supply. Although infiltration systems will not prevent large flood events, the reduced runoff can help to reduce hydraulic loading of drainage systems. This can contribute towards the alleviation of flood problems and may also result in other benefits such as the reduced operation of combined sewer overflows.

Permeable surfaces may be particularly appropriate for low-loading traffic areas (such as car parks). In addition to permeable surfacing (see Figure 7.2) and various types of infiltration facilities (see Figure 7.3), there are various types of soakaway devices including perforated pipes, basins, trenches and swales.

The main operational constraint for these technologies is the potential problem of clogging due to poor solid waste management, poor street sweeping and a lack of maintenance. Other potential problems relate to the risks of groundwater pollution. Potential conflicts between the benefits of groundwater recharge and the risks to long-term groundwater quality need to be considered carefully. In addition, infiltration facilities should not be located near to septic tanks or building foundations. The use of infiltration is therefore generally considered to be feasible in areas where there is considered to be a low risk of groundwater contamination, soils with good infiltration rates and deeper groundwater or bedrock.

Many of these technologies have yet to be widely adopted but there is increasing interest in both developed and developing countries for the potential of these systems.

112 Urban stormwater management in developing countries

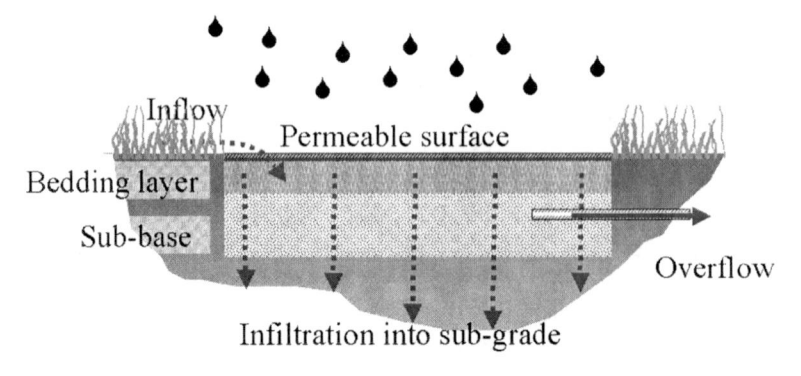

Figure 7.2 Permeable surface (Martin *et al.* 2000).

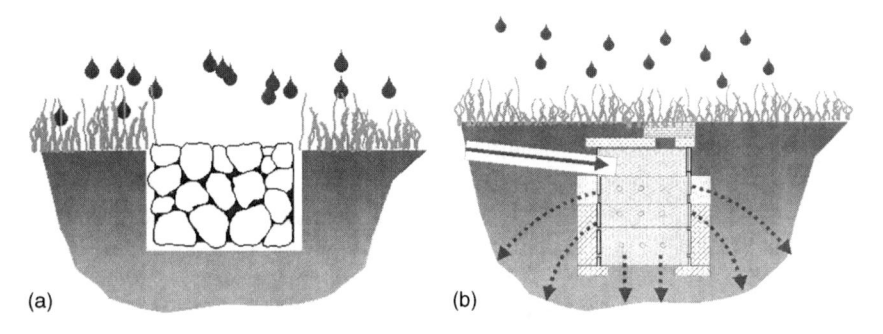

Figure 7.3 Infiltration (a) pit and (b) soakaway (Martin *et al.* 2000).

Box 7.3 describes experience from Chile in the development of permeable surfacing and stormwater infiltration wells and Figure 7.4 shows the application of these infiltration wells in a new housing development.

7.4 SWALES

Swales are vegetated, open channels that have a dual function to control runoff. In addition to the attenuation of peak flows of stormwater runoff via storage, the permeable base and sides of the channel promote infiltration. The vegetation also reduces flow velocity in the channel due to high roughness and contributes towards reduction of pollutant concentrations in the runoff. Where the permeability of the soil is low, the swale may be combined with a perforated pipe system in which a shallow grass swale is underlain by a continuous section of perforated pipe enclosed in a trench filled with granular material. Grassed swales are increasingly

Box 7.3 Development of prototype stormwater infiltration technologies in Chile

In 1997, after many years of a lack of interest in urban drainage but increasing frequency of urban flooding problems, particularly in the larger urban centres, a new Stormwater Act was published in Chile. This new regulation requires the construction of stormwater mitigation devices in all new developments. To contribute to the solution of flooding problems, Pontificia Catholic University of Chile developed a prototype design for permeable pavement and infiltration wells.

The houses of a new estate were installed with small infiltration wells (1 m deep by 0.8 m diameter), which were connected directly to the roof drainage. The overflow from each of these infiltration wells is connected to the collector drainage system. To function properly the wells must receive water directly from the roof, and each device was fitted with a leaf and debris separator in order to minimise clogging. Based on 2 years of rainfall-runoff data, compared with the normal runoff situation, the houses fitted with the infiltration wells showed a 10–20% reduction in the mean value of stormwater volume and, more importantly, a 20–40% reduction in the stormwater peak.

In addition to the experiments with infiltration wells, a concrete mix has been developed and used in the construction of a parking lot with a permeable surface. Observations verify the large infiltration and storage capacity of the layers of the permeable surface. No surface runoff was observed from any storm event occurring during the 2-year period of monitoring. In addition, the subsurface outflow hydrographs demonstrated the considerable attenuation capacity, particularly for the intense storms of short duration. Given the great permeability of the concrete mix and of the sub-base gravel layers, the infiltration capacity of the pavement is only limited by the natural soil permeability. Thus, the system is also of interest in soils of low permeability since the storage capacity of the gravel layer allows stormwater to be retained and drained slowly to the downstream drainage network.

Source: Professor Bonifacio Fernández,
Pontificia Catholic University of Chile

being employed in developed countries in temperate climates as flow and quality control measure. However, they have potential application in many parts of the world and, as shown in Figure 7.5, are appropriate for low-density developments with soils with good infiltration capacity and low ground-water tables. As shown in Figure 7.6, there is also opportunity to combine swales with infiltration trenches to increase the efficiency of the swale by promoting infiltration capacity.

As well as runoff control, grass swales show good performance for removal of large particles such as suspended solids and pollutants attached to solids (e.g. phosphorus). The primary mechanisms for pollutant removal in swales are settling

Figure 7.4 Houses provided with small infiltration wells in Chile. (Photo: Bonifacio Fernández.)

Figure 7.5 A swale on the engineering campus at Universiti Sains Malaysia (USM) (Abdullah *et al.* 2004).

and adsorption of particulates. In addition, particulates are filtered and dissolved solids are absorbed by vegetation. However, pollutant removal efficiencies of swales vary widely and further research is required to evaluate the performance of swales in both temperate, but particularly in tropical climates under the conditions of high intensity of rainfall (Mohd Sidek *et al.* 2002; Zakaria *et al.* 2004).

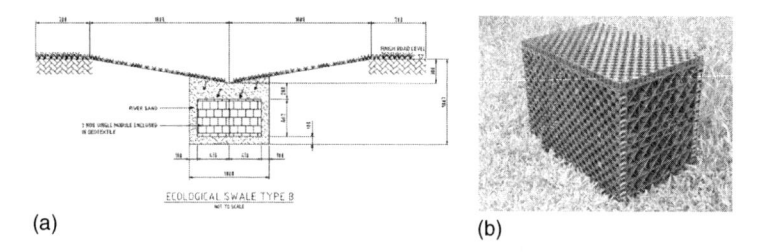

Figure 7.6 (a) A swale combined with an infiltration trench. (b) The media used for the trench (*Source*: REDAC, USM).

7.5 CONSTRUCTED WETLANDS

Constructed wetlands are increasingly being used to manage urban stormwater with the objectives of runoff control and water quality improvement. Constructed wetlands also have landscape and amenity value, improving the aesthetical value of the urban environment and providing a natural habitat for flora and fauna.

There are a number of different types of design of wetland but, due to the high flow rates and relatively low concentration of pollutants in stormwater (compared with other wastewaters), constructed wetlands for urban drainage are generally designed for horizontal surface flows. Constructed wetlands are planted with vegetation to improve pollutant removal mechanisms by various processes, including filtration, infiltration and biosorption. They are effective in removing particulates and dissolved pollutants, nutrients as well as toxic substances such as heavy metals. However, performance is dependent on the concentration and characteristics of the pollutants, and especially the association of these pollutants with different particle size fractions of stormwater sediments.

The interaction of hydraulic, physical and biological factors directly determines the treatment performance of constructed wetlands. Wong *et al.* (1999) discuss the various issues and performance considerations associated with wetlands for stormwater pollution control. The most important design consideration is the short-intermittent nature of storm runoff. Due to this variability, important design criteria for particular wetland features may vary depending on site-specific characteristics. The detention time is commonly used to estimate the performance of these facilities as pollution control systems. However, it is widely recognised that other factors such as the flow hydrodynamics within the detention system and vegetation density and layout can have a significant influence on their performance.

Drainage engineers considering the use of constructed wetlands might be concerned that mosquitoes could become a nuisance and potentially dangerous in areas where mosquito related diseases are prevalent (see Chapter 2). Although this is a real concern that cannot be ignored, the problems may not be as significant as feared. Wetlands can be designed so as to deter mosquito breeding by ensuring that the residence time is only a few days so that the larvae do not have time to grow. In addition, various biological controls can be employed to control the insect population and these are preferred over chemical controls, since biological controls are generally cheaper and pose less risk to humans (McLean 2000; Walton 2003).

In general, wetlands need little attention to operational and maintenance requirements, but the reeds need to be cut every year and the reeds need to be disposed off or reused. However, due to a lack of effective management structures and a lack of understanding of the natural wetland functions, maintenance is generally poor, particularly in developing countries. Therefore, where new technologies are introduced, it is important that they are accompanied with capacity building of local institutions to ensure that they are adequately monitored and maintained. As described below, this can be combined with demonstration projects, which can also be used to highlight a wider awareness about the benefits of these technologies.

7.6 PRACTICAL DEMONSTRATION PROJECTS OF SUSTAINABLE URBAN DRAINAGE

The technologies described in this chapter provide a significant departure from the conventional approach towards urban drainage and flood control. As a result of the increasing interest in these approaches, there is a tendency to develop or re-develop stormwater infiltration facilities and, in some countries (e.g. Switzerland), infiltration of stormwater is mandatory for new developments. Other countries such as Germany, Sweden and Denmark in Europe as well as Australia and USA are actively encouraging the use of various source control technologies and other best management practices through local or municipal incentive measures (Maršálek and Chocat 2002). There is also increasing interest in many developing countries such as Brazil, Chile and, as described below, in Malaysia.

Practical examples and demonstration projects of source control are an important and necessary part of the development and promotion of these new approaches. A number of programmes have been instigated in various countries such as an on-going project by the Construction Industry Research Information Association in the UK (Martin *et al.* 2000; Wilson *et al.* 2004). Dallmer (2002) describes another programme undertaken by South Sydney City Council in Australia to promote sustainable water management and assess the potential for on-site stormwater treatment and reuse. The aim of the project, named Stormwater Quality Improvement and Reuse Treatment Scheme (SQIRTS) is to demonstrate best-practice water management approaches, to learn from the process of implementing and to encourage the further use of these technologies. With the demonstration of stormwater reuse principles within SQIRTS, the City Council has begun to promote an increase in reuse projects throughout the South Sydney area. As a result of the project, there are planned amendments to Council's policy to ensure that new developments address reuse as one action within a suite of sustainability initiatives.

In the following section, another example of a pilot project and how it has been used to demonstrate principles of best practice in Malaysia is described. The experiences from Malaysia are also very relevant in many other parts of the world, and other countries considering the adoption of similar strategies for management of urban runoff could learn from these experiences.

Case study: The Bio-Ecological Drainage System project, Malaysia

The developments in Malaysia offer an exemplary model for urban stormwater management for other countries in the region and other parts of the world with similar tropical climates. The Bio-Ecological Drainage System (BIOECODS), a pilot project conceptualised and designed by researchers from River Engineering and Urban Drainage Research Centre (REDAC), Universiti Sains Malaysia (USM) is an innovative drainage system based on the 'Urban Storm Water Management Manual' (Zakaria *et al.* 2004) approved by the Department of Irrigation and Drainage in January 2001 to replace previous planning and design procedures for urban drainage systems in Malaysia.

The BIOECODS project is implemented at USM's Engineering Campus and has following objectives (Mohd Sidek *et al.* 2002; Zakaria *et al.* 2003; Ab. Ghani *et al.* 2004):

(1) To offer a demonstration project for the use of new ecological drainage systems for private and public buildings.
(2) To assess the performance, potential application and continued use of integrated ecological drainage systems under Malaysian conditions.
(3) To monitor and study the effectiveness of individual technologies in terms of runoff and pollution control.
(4) To develop a modelling procedure for the analysis, design and optimisation of integrated ecological drainage systems.
(5) To evaluate the cost-effectiveness of integrated ecological drainage systems.
(6) To provide guidelines for new ecological drainage systems for use in Malaysia.

The main function is to promote stormwater infiltration from impermeable areas (e.g. roof tops, car parks, etc.) by using bio-ecological swales. The second function is to gradually release the stormwater through the use of swales, underground detention storage and dry ponds. The third function is to improve receiving water quality by the use of technologies to treat stormwater using a series of technologies including swales, wet ponds and wetlands.

The schematic diagram of BIOECODS drainage system for USM Engineering Campus is shown in Figure 7.7, and is divided into seven steps before the runoff reaches the receiving water.

The results of the study indicate that the BIOECODS can be a viable method for the water quantity and quality treatment of site runoff. Preliminary data

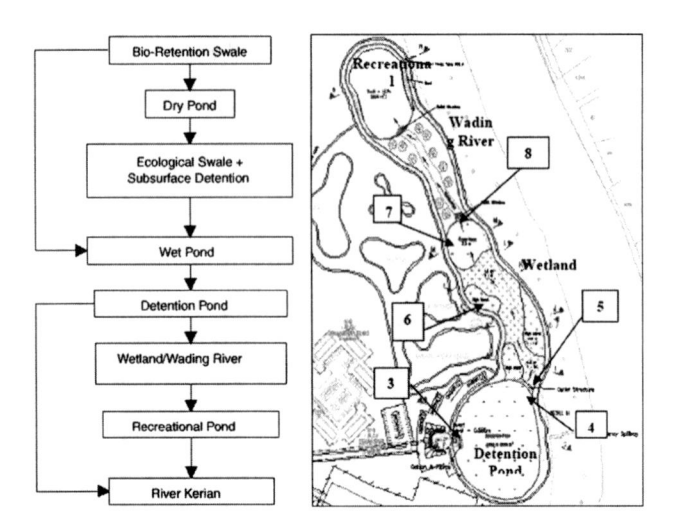

Figure 7.7 Schematic diagram and layout of the BIOECODS on the engineering campus at USM (Ab. Ghani *et al.* 2004; Mohd Sidek *et al.* 2004).

collection between June and December 2003 indicates the following results:

- The ecological swale performance shows that the reduction in peak flow ranges from 30% to 56% for surface swale while for subsurface infiltration trench up to a maximum of 60% (Ainan *et al.* 2004).
- Water quality samplings at 10 points for different type of swales show that most parameters such as dissolved oxygen (DO), biological oxygen demand (BOD), pH and total suspended solids (TSS) comply with Class II, National Interim Water Quality Standard for Malaysia (Mohd Sidek *et al.* 2004).
- Water quality results for the ecological pond system made up of a wet pond, a detention pond and a wetland show that the treated stormwater runoff falls into Class II, National Interim Water Quality Standard for Malaysia (Mohd Sidek *et al.* 2004).

Based on the experience of the construction of BIOECODS at USM's Engineering Campus, several further considerations are proposed for the construction of similar drainage systems in other locations. There is a need of paradigm shift in Malaysian way of thinking and action towards sustainable drainage systems. For this to be possible, a public campaign needs to be conducted nationwide to highlight the new concept of drainage system. The very important aspect to ensure the successful of the stormwater programme is to educate the public (public awareness). For any new development, the partnership between the developer and responsible authority should be encouraged through involvement in the demonstration project. The developer must agree to incorporate sediment control systems, innovative stormwater control systems and design concepts that minimise land-disturbing activities during construction. More specifically, the developer should be responsible for providing the land, meeting the capital costs and landscaping costs of installation of these technologies. However, legal arguments about who is responsible for the long-term maintenance are likely and it will be necessary to try to resolve these conflicts through a process of discussion, mediation and arbitration.

7.7 REFERENCES

Ab. Ghani, A., Zakaria, N. A., Abdullah, R., Yusof, M. F., Mohd Sidek, L., Kassim, A. H. and Ainan, A. (2004) Bio-ecological drainage system (BIOECODS): concept, design and construction. *Proceedings of the 6th International Conference on Hydroscience and Engineering (ICHE-2004)*, Brisbane, Australia, 30 May to 3 June.

Ainan, A., Zakaria, N. A., Ab. Ghani, A., Abdullah, R., Mohd Sidek, L., Yusof, M. F. and Wong, L. P. (2004) Peak flow attenuation using ecological swale and dry pond. *The 6th International Conference on Hydroscience and Engineering (ICHE-2004)*, Brisbane, Australia, 30 May to 3 June.

Dallmer, L. (2002) SQIRTS – an on-site stormwater treatment and reuse approach to sustainable water management in Sydney. *Water Science and Technology* **46**(6–7), 151–158.

Dhar Chakrabarti, P. G. (2001) Rooftop rainwater harvesting as an alternative technology for fresh water augmentation in chronically deficient urban agglomerates of India. *Proceedings of a Symposium 'Frontiers in Urban Water Management: Deadlock or Hope?'*, Marseille, France, 18–20 June 2001. In: *IHP-V Technical Documents in Hydrology No. 45* (eds. José Alberto Tejada-Guibert and Čedo Maksimović), UNESCO, International Hydrological Programme, Paris, pp. 191–199.

DTU (2002) *Very-Low-Cost Domestic Roofwater Harvesting in the Humid Tropics: Constraints and Problems.* Development Technology Unit, Domestic Roofwater Harvesting Research Programme. (DFID KaR Contract R7833 Report R20), School of Engineering, University of Warwick, UK.

Girling, C., Kellett, R., Rochefort, J. and Roe, C. (2000) *Green Neighborhoods – Planning and Design Guidelines for Air, Water and Urban Forest Quality*. Center for Housing Innovation, University of Oregon, USA.

Gould, J. (2000) Rainwater catchment systems: reflections and prospects. *Waterlines*, **18**(3).

Maršálek, J. and Chocat, B. (2002) International report: stormwater management. *Water Science and Technology* **46**(6–7), 1–17.

Martin et al. (2000) *Sustainable Urban Drainage Systems, Best Practice Manual*. CIRIA Report 523, Construction Industry Research Information Association, London.

McLean, J. (2000) Mosquitoes in constructed wetlands: a management bugaboo? *Article 100 in The Practice of Watershed Protection*, Center for Watershed Protection, Ellicott City, USA.

Mitchell, G., Mein, R. and McMahon, T. (1999) *The Reuse Potential of Urban Stormwater and Wastewater*. Industry Report 99/14. Co-operative Research Centre for Catchment Hydrology. Industry Report 97/11. Monash University, Australia.

Mohd Sidek, L., Takara, K., Ab. Ghani, A., Zakaria, N. A. and Abdullah, R. (2002) Bio-ecological drainage systems (BIOCEDS): an integrated approach for urban water environmental planning. *Seminar on Water Environmental Planning: Technologies of Water Resources Management*, Kuala Lumpur, 15–16 October 2002.

Mohd Sidek, L., Ainan, A., Zakaria, N. A., Ab. Ghani, A., Abdullah, R. and Ayub, K. R. (2004) Stormwater purification capability of BIOECODS. *The 6th International Conference on Hydroscience and Engineering (ICHE-2004)*, Brisbane, Australia, 30 May to 3 June.

Vishwanath, S. (2001) Domestic rainwater harvesting – some applications in Bangalore, India. *International Conference on Rainwater Harvesting*, New Delhi, April 2001, IITD. www.rainwaterclub.org

Walton, W. E. (2003) *Managing Mosquitoes in Surface-flow Constructed Treatment Wetlands*. Publication 8117. Division of Agriculture and Natural Resources, University of California. Vector-Borne Disease Section, California Department of Health Services. http://anrcatalog.ucdavis.edu.

Wehrle, K. (2001) Securing domestic water supplies for rural Bangladesh through combined systems. *International Rainwater Catchment Systems Association, 10th International Conference on Rainwater Catchment Systems 'Rainwater International 2001'*, Mannheim, Germany, September.

Wilson, S., Bray, R. and Cooper, P. (2004) *Sustainable Drainage Systems. Hydraulic, Structural and Water Quality Advice.* CIRIA Report 609, Construction Industry Research Information Association, London.

Wong, T. H. F., Breen, P. F. and Somes, N. L. G. (1999) *Ponds vs Wetlands – Performance Considerations in Stormwater Quality Management.* Co-operative Research Centre for Catchment Hydrology, CRCCH Technical Report. Monash University, Australia.

Zakaria, N. A., Ab. Ghani, A., Abdullah, R., Mohd Sidek, L. and Ainan, A. (2003) Bio-Ecological Drainage System (BIOECODS) for water quantity and quality control. *International Journal River Basin Management*, IAHR & INBO, 1(3), 237–251.

Zakaria, N. A., Ab. Ghani A., Abdullah, R., Sidek, L. M., Kassim, A. H. and Ainan, A. (2004) MSMA – A new urban stormwater management manual for Malaysia. *Proceedings of the 6th International Conference on Hydroscience and Engineering (ICHE-2004)*, Brisbane, Australia, 30 May to 3 June.

8
Applications of computer models

The management of water in urban areas is complicated and costly. Any construction or modification to urban drainage infrastructure needs to be well planned so that its impact is known in advance. Computer models provide planners and engineers with versatile tools to help understand the hydrological and hydrodynamic processes in a catchment. They can also be used to evaluate the impacts of future development scenarios and assess the implications of possible changes to the existing situation and/or proposed physical interventions. Although the development of a computer model can require considerable investment in terms of time and resources, the long-term benefits can be significant if the model is used to develop a cost-effective solution to urban drainage problems.

8.1 COMPUTER MODELLING FOR HYDROLOGY AND HYDRAULICS

8.1.1 What is a model?

A model is basically a simplified representation of a part of the real world that may be used to make detailed analyses, which would otherwise not be feasible in the real world situation. There are two main types of models that may be applied for the investigation of hydrological and hydrodynamic processes in urban catchments.

© 2005 IWA Publishing. *Urban Stormwater Management in Developing Countries* by Jonathan Parkinson and Ole Mark. ISBN: 1843390574. Published by IWA Publishing, London, UK.

8.1.1.1 Physical models

Traditionally, physical models built to a small scale (1 : 25 to 1 : 100) in a hydraulic laboratory[1] have been used to simulate hydraulic structures such as dams, harbours, rivers or drainage systems (see Figure 8.1). These physical models enable engineers to evaluate their design in detail and to predict the performance of the actual structures before they are built.

Figure 8.1 Physical model of a sewer with a manhole at the University of Sheffield, UK.

---

[1] The downscaling must be in accordance with the appropriate similarity law, that is the law of Froude (or Reynolds) in order to encapsulate the ratio between gravity (or viscous) forces and inertial forces of the real world in the downscaled model.

8.1.1.2 *Mathematical models*

With the arrival of computer modelling in the late 1960s, it became possible to solve the complex deterministic mathematical equations used to describe flows and water levels in hydrological and hydraulic systems. This provided engineers with a powerful tool to test and evaluate the performance of these systems without having to build them in the laboratory. The arrival of more powerful computers enabled mathematical models to advance rapidly. Over the past 30 years, an increasing number of computer models have been developed and applied to assist in the planning and design of urban drainage systems.

Like physical models, mathematical models are still simplified representations of the real world. However, computer modelling has reached a stage that enables highly complex systems to be simulated, which goes far beyond what is possible to achieve using physical models. In some cases, for example, for the cities of Bangkok and Buenos Aires, the complete sewerage and drainage systems are managed with the aid of a computer model. In these situations, the models are very complex and consist of vast amounts of data describing the different surface types and thousands of pipes and drainage channels.

8.1.2 Types of mathematical model

There are two types of mathematical models that are applied for design and simulation of urban drainage and stormwater systems. These are either based on a deterministic approach, that is, with a fixed relationship between the model input and its results, or a stochastic approach, which generates series of model results with given statistical properties. Models for urban drainage and stormwater are predominantly deterministic models, and this chapter refers only to these.

Deterministic models applied for urban drainage can be divided into the following two classes:

(1) Conceptual models rely on empirical parameters that need to be calibrated by comparing the model output to field measurements. The majority of surface runoff models are conceptual models.
(2) Physically-based models solve the equations describing fundamental physical relationships based on parameters which are usually derived from observed measurements. However, some parameters still have to be calibrated. Hydrodynamic network models, which solve the Saint Venant equations,[2] are examples of physically-based models.

The majority of urban drainage software incorporate both conceptual models for the simulation of surface runoff and physically-based models for the simulation of flows in the drainage system itself.

---

[2] The Saint Venant equations are applicable for one-dimension flow, with a mild slope and a small streamline curvature.

8.1.3 Applications of models for urban drainage and stormwater systems

Urban drainage and stormwater models are usually setup and applied to perform specific tasks including:

(1) Design of new systems.
(2) Analysis of existing systems – identification of flooding and urban drainage problems.
(3) Analyses of upgrading measures with the objective to optimise the existing system – this may include rehabilitation of the drainage system, or improved control strategies for pumps, weirs and gates.
(4) Provision of information for operation and real-time decision support (e.g. for flood-response strategies).
(5) To provide input to economic analyses of the impacts of flooding.
(6) Analyse flooding scenarios to assist in the development of flood-response strategies.

The applications described above traditionally deal with flood control and pollution control. These analyses may contain information such as:

- Characterisation of urban runoff with respect to temporal and spatial flow distributions and pollutant concentration/load ranges.
- Input to a wastewater treatment plant and/or receiving water quality analysis.
- Determination of effects and relative benefits of various options for control of runoff – either singularly or in combination.
- Frequency analysis on hydrological or quality parameters (e.g. to determine return periods of flows/concentration/loads).

The next sections describe in more detail the most common applications of urban drainage and stormwater models.

8.1.3.1 Pollution control

Modelling of wastewater is important for predicting the dry weather inflow (e.g. to a stream, a wetland or a wastewater treatment plant). It is also important for accurate descriptions of overflow from sewer systems. Traditionally, modelling of overflows for combined sewer systems is carried out using a rainfall-runoff model as input to a hydrodynamic flow model in conjunction with a diurnal description of the wastewater. The pollutant load to streams and rivers is typically computed using a simplified hydraulic model in combination with a plug-flow-pollution model, for example SAMBA (Johansen et al. 1984) or SIMPOL (Foundation for Water Research 1998).

These kinds of models might include computation of a time-area runoff hydrograph together with the most significant elements in the pipe/drainage system (i.e. weirs, pumps, locations with divergent flow and outlets). Due to the simplifications, the calculations are very fast and suitable for assessing large numbers of alternative layouts of drainage systems. In the very early days of computer models this type of modelling was the only way to calculate annual loads of discharges of water and pollutants based on historical time series of rainfall. However, the development of more advanced hardware and software means that complex simulations can be carried out much more rapidly using more advanced computer models.

8.1.3.2 Flood control

In the modelling of wastewater systems, the flows are confined to pipe or drains with well-defined dimensions, and can be measured accurately and described by the one-dimensional Saint Venant equations. Modelling of urban flooding is fundamentally different as the runoff flows over a three-dimensional surface, where the flow routes and paths are not restricted in an obvious manner. Therefore, the modelling of the surface flooding requires additional data in an environment where the local geometry is more difficult to describe accurately. The best approach is to describe the surface using a digital elevation model containing roads and houses in the flooded areas.

Modelling of small-scale local flood problems has been around for the last 30 years, and traditional flood control modelling applications have been used to:

(1) Identify the cause of flooding.
(2) Determine design of embankments based on a flood return frequency.
(3) Determine pumping capacities (when applicable).
(4) Design ponds, basins, etc. for flood storage.

Modelling of large-scale urban flooding in situations – where a major part of a city is flooded – is an emerging area where practical applications have started to appear only during the last decade. An example of this from Dhaka is described below.

8.2 THE MODELLING PROCEDURE

The development of models to address practical engineering problems involves the following steps:

(1) Planning and preparation (including an assessment of data requirements).
(2) Acquire data, formulate and build model.
(3) Model calibration and validation.
(4) Evaluate the performance of the calibration and validation. If necessary repeat Steps 2 and 3 to improve the model accuracy if it is not considered to be satisfactory.
(5) Model application and assessment of results.

These steps are discussed in the following sections and illustrated in Figure 8.2.

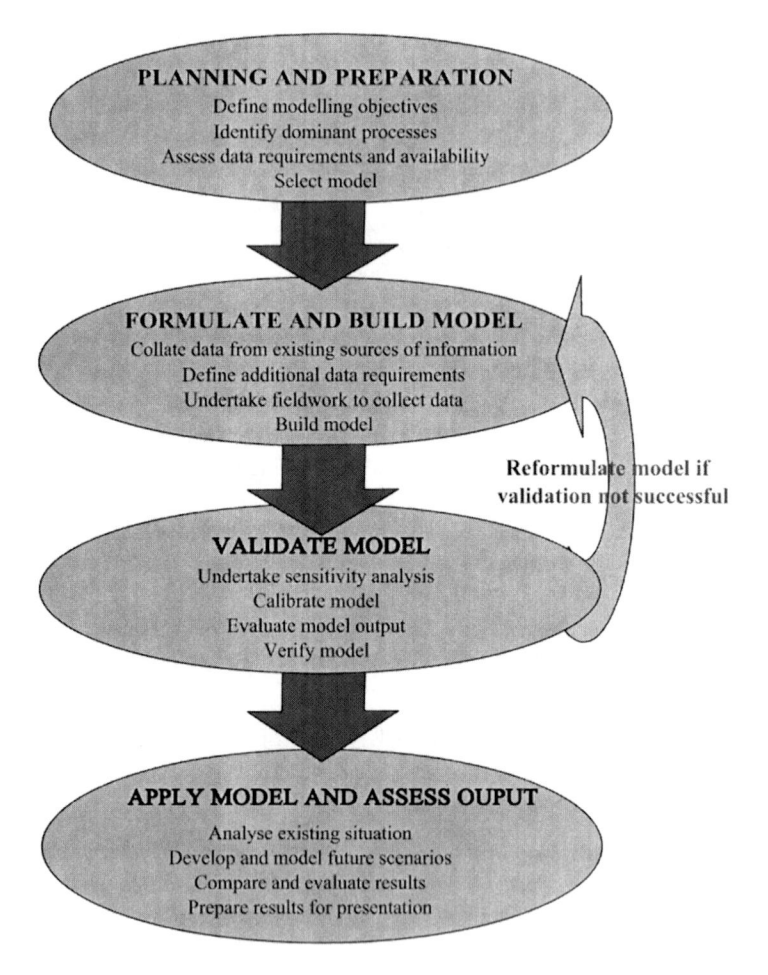

PLANNING AND PREPARATION
Define modelling objectives
Identify dominant processes
Assess data requirements and availability
Select model

FORMULATE AND BUILD MODEL
Collate data from existing sources of information
Define additional data requirements
Undertake fieldwork to collect data
Build model

Reformulate model if validation not successful

VALIDATE MODEL
Undertake sensitivity analysis
Calibrate model
Evaluate model output
Verify model

APPLY MODEL AND ASSESS OUPUT
Analyse existing situation
Develop and model future scenarios
Compare and evaluate results
Prepare results for presentation

Figure 8.2 Steps in the modelling procedure.

8.2.1 Planning and preparation

8.2.1.1 Objectives and dominant processes

During the planning phase of any project in which a model is to be used, the objective of the modelling must be clearly defined. The two most common aims are to reduce flooding and to reduce pollution spills from the drainage system into receiving waters.

When the purpose has been defined clearly, the dominant physical, chemical and biological processes involved should be identified. Based on these considerations, together with a specification of the accuracy of the model output, a model should be selected.

8.2.1.2 Assessing data requirements

The data requirements of the model must be compared with the availability of data and the following questions should be considered:

(1) Are available data sufficient to describe the processes involved in the modelling?
(2) Is it feasible to collect the data needed for the analyses during the project?
(3) Does the model need specific boundary conditions for example in the form of a downstream water level?

In some cases it is not possible to carry out the amount of data collection and measurements required for the modelling – it is simply too expensive or the duration of the project is too short. In this case, it is necessary to evaluate whether or not to proceed with the modelling – and to assess if a less data intensive modelling approach would be a suitable alternative.

8.2.1.3 Model selection

The model selected for the purpose must include the dominant processes involved in the study and it must have input requirements corresponding to the availability of data. It is not necessarily an advantage to apply an advanced, physically-based model rather than a simpler tool. For example, in basic rainfall-runoff modelling, where only one or two dominant physical processes are present, simple models can produce runoff simulations of similar accuracy as advanced models even though their data requirements are much lower and they are easier and faster to use.

However, whichever type of model is used, in order to carry out a successful modelling exercise, it is necessary to ensure that resources are available for the acquisition of the hardware and software necessary to undertake the work. In addition, human skills are required to understand the limitation of the available models, and to understand and analyse the model output. Without a proper understanding of the physical environment involved, modelling becomes a risky business, where too many resources are spent without achieving the desired goal.

8.2.2 Data acquisition and model building

Once the most appropriate model has been selected, although it may be laborious, it is generally relatively straightforward to build the model based on the input data requirements. Reliable field measurements of flows and concentrations of wastewater in sewers and drains are essential for successful model validation. If some local data are unavailable, the relevant calibration parameters may be obtained from literature or other projects with similar conditions in which the same model was used. However, it is very important to note that using default values, literature values or prior experience from similar catchments will decrease the accuracy of the model prediction.

As described in Chapter 5, many software packages enable direct links to Geographical Information Systems (GIS) and, in this way, data concerning the catchment and the drainage system can be extracted from the GIS system and automatically fed into the data structure of the selected model. However, after importing data from a GIS system, the model data must be checked very carefully as GIS data often contains errors (original digitising errors or a lack of connectivity in the drainage system data), which may be fatal for the modelling exercise. Furthermore, there are many data requirements that are generally not available from a GIS and specific model data (e.g. Manning numbers to describe drainage conduit roughness) must therefore be estimated and verified as part of the model calibration. This data will need to be collected separately and the data must be entered manually into the model. This may be time consuming, but a necessary and important part of the modelling procedure.

8.3 MODEL CALIBRATION AND VALIDATION

8.3.1 Sensitivity analysis

The capabilities and nuances of a model should be investigated through a sensitivity analysis, which involves observing the effect and importance of each parameter by systematically changing its value and assessing the impact on the model results. The objective of this exercise is to identify the parameters that need to be more carefully determined. Sensitivity analyses also help to describe the corresponding need for data accuracy. For example, the required temporal resolution of the input time series, such as rainfall, may be assessed by applying data of different spatial and temporal resolution, and comparing the results.

Test runs with the model at this stage will show if the available data in conjunction with the selected model provides results that are deemed to be of sufficient quality. If the results are unsatisfactory then the adopted modelling approach (including the applicability of the selected model) must be reassessed.

8.3.2 Calibration

Urban drainage models are expected to simulate the behaviour of the modelled system with a reasonable level of accuracy. This is usually ensured through the process of parameter calibrations and the validation of model results against the measured performance of the real system. Calibration involves adjustment of the model parameters, with the objective of minimising differences between the simulated model results and the observed field measurements (e.g. of water levels). This is normally a manual iterative procedure but some software packages include automatic calibration routines, which may provide a fast first estimate of the calibration parameters. However, as the calibration procedure may find a solution that is not optimal for the chosen modelling objective (see below about the selection of the objective functions), automatic calibration results should be checked by someone with a fair understanding of model parameters.

The focus of the model calibration exercise is not the same for all types of models. By their nature, conceptual models require more attention in this respect than the physically-based models. Therefore, in urban drainage modelling, the focus is usually on the hydrological model (conceptual) component, while the deterministic hydrodynamic model component of the drainage network typically requires only minor adjustments for accurate performance. However, independently of the applied model type, good modelling practice should involve thorough model validation before application.

A continuous period or a set of intermittent events used for calibration should preferably include the full range of expected operational conditions and natural variations in the system. To calibrate urban drainage models, the following objective functions may be considered:

(1) Correlation between the simulated and observed catchment runoff volume (i.e. a good water balance).
(2) Overall agreement of the shape of the rainfall-runoff response hydrograph.
(3) Correlation between the peak flows with respect to timing, rate and volume.
(4) Description of the dry weather flows in order to describe the water quality of the runoff during storm events.

In general, trade-offs exist between these different objective functions. For instance, one set of parameters may offer a very good calibration in relation to simulation of peak flows but poor calibration in relation to low flows, or vice versa. In practical applications, the user can select any of the four objective functions or a combination, depending on the purpose of the specific model application. For example, the overall shape of the hydrograph is relevant for studying flooding problems, whereas agreement of runoff volumes can be more important for the design of stormwater detention basins.

The calibration of a model involves changing parameters until it simulates the observed measurements of the real life processes that it is intended to represent. The constraints of calibrating a mathematical model are illustrated in Figure 8.3. The 'model domain' represents the catchment that is to be modelled. The physical system (a catchment in this case) is shown on the left and the mathematical model is shown on the right. The simplification of the catchment and the associated physical processes inevitably leads to errors in the model structure when the model simplifies the real life. The model input and output parameters are quantified by measurements that contain some uncertainty, in terms of both quantity and timing. The model produces output containing the effects of the input deficiencies. This simulation output is then compared with measurements, which have themselves other uncertainties due to the way the measurements were carried out.

Differences between observations and simulated model response can therefore originate from one or more of the following five different error sources:

(1) Input model data.
(2) Recorded observations (e.g. flow and water levels) used for calibration of the model.

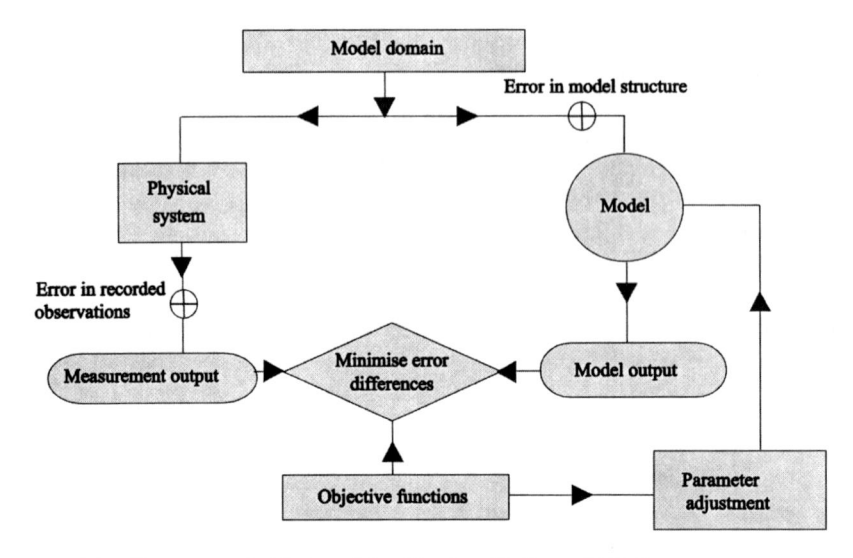

Figure 8.3 The concept behind model calibration (Refsgaard 1995).

(3) Non-optimal parameter values in the model.
(4) Anomalies inherent in the model (e.g. such as numerical dispersion from the solution of differential equations).
(5) Invalid model structure.

Disagreements between simulated output and observed input data can be due to any of the five error sources listed above, but only error source (3) can be minimised during model calibration. The measurement errors related to error source (2) should generally only be 'background noise' and give a minimum level of disagreement below which further parameter or model adjustments will not improve the results significantly.

The primary objective of the calibration procedure should therefore be to reduce the error source (3) to a level to which it becomes insignificant compared with the data error sources (1) and (2). However, if the errors sources in (1) and (2) are considered to be unacceptable large – then either a new model structure has to be chosen or an additional measurement exercise must be undertaken to collect more data for further calibration.

Thus, it is important that a clear distinction is drawn between the different error sources during the calibration process, so that compensation for errors from one source by adjustments within another source is not attempted. Otherwise the calibration will degenerate to curve fitting (e.g. if an error in the measurement of the ground slope is compensated by applying a Manning number which is beyond physically meaningful limits). Curve fitting may result in a reasonable calibration based on the observed data but will inevitably give poor predictions when the model is applied to simulate other data sets.

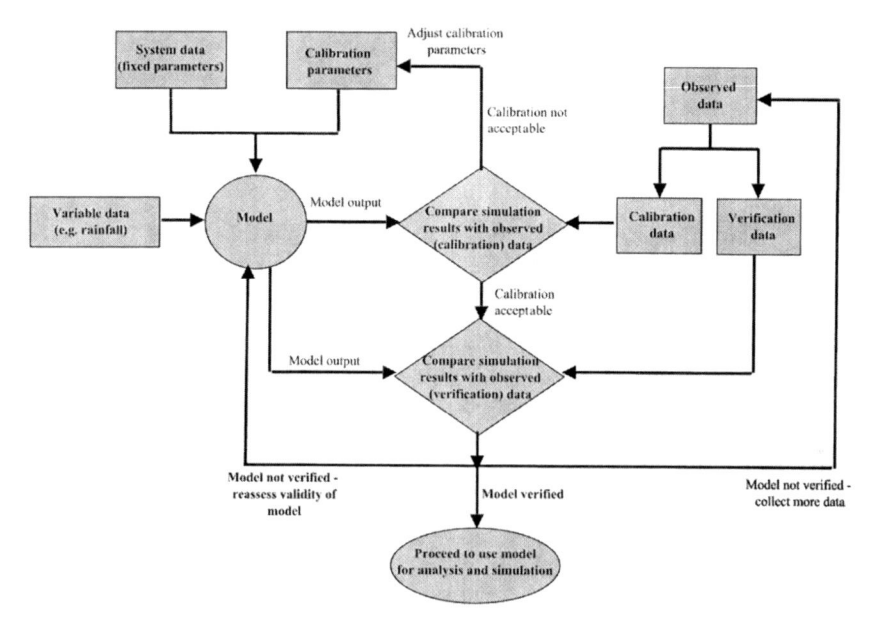

Figure 8.4 The calibration and validation procedure.

8.3.3 Model validation

Before an urban drainage model is used as a basis for any decisions, for example design of upgrading measures on an existing system, the performance of the model simulations must be checked to ensure that it accords with actual system performance. The workflow of the calibration and validation procedure can be seen in Figure 8.4.

The purpose of model validation is to provide evidence that the model generates results within an acceptable error range before it is applied using a new set of observed data that was not used for the calibration. Hence, validation must be carried out for a simulation period or intermittent event(s), which are different and independent of those used for the model calibration. The results of these simulations may be less accurate than the calibration results, particularly if the input data go beyond the range of data applied in the calibration.

Hence, it is important to recognise that the performance of a model during calibration may not be representative of the overall performance of the model using a wider data set. Special caution must be taken if a model is applied for input data, which is an extension of the range covered by the calibration data. If a model does not pass the validation procedure, then it must be reassessed according to the following:

- If more data should be collected.
- If the model structure is suitable for the purpose.

- If another model would be more appropriate.
- If there is good reason to stop the modelling exercise altogether if it is felt that it is not possible to achieve satisfactory results.

8.4 MODEL APPLICATION

8.4.1 Design storm events and performance analysis

For design purposes or for performance analysis, various rainfall data sets of different duration are used to evaluate and compare the performance of the system under different storm conditions. This rain data can also be used to assess various proposals for structural modifications or extensions to the drainage system. Potential structural changes can then be carried out by adding these to the model. The impact of the proposed measures can be analysed, for example, aiming at improving the discharge capacity of the existing network through increased dimensions of the canals, increased pumping capacity or through the provision of attenuation ponds (detention or retention ponds) for the storage of peak runoff to reduce downstream flood events (see Chapter 6).

These rainfall events (otherwise referred to as design storms) may consist of synthetically generated rainfall or using recorded data from observed rain events. Various synthetic design storms (for a certain catchment) of a 10-year return period with durations of 30, 60 and 180 minutes are shown in Figure 8.5.

Using design storm events or recorded data as input data for analysis of the system provides the basis for evaluation of the drainage network performance. The most critical parts in the system can be identified, that is where flooding is most

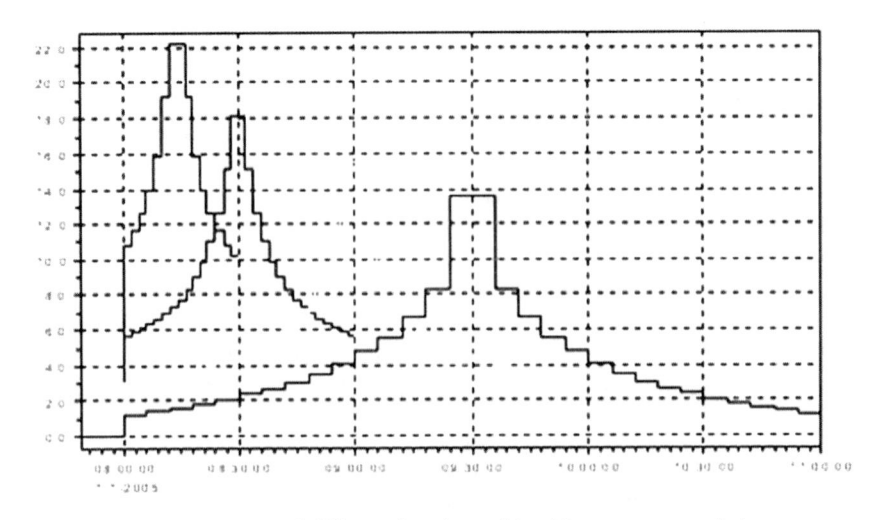

Figure 8.5 Design storms of different durations with a 10-year return period.

likely to occur or where discharges of polluted runoff are considered to cause a problem in receiving waters. Two examples of model application for flood control are described below.

Case study: Urban stormwater drainage system in Addis Ababa, Ethiopia – present state and proposals for improvement

The sewerage system in Addis Ababa, the capital of Ethiopia serves a population of 200,000, which is less than 10% of the total population of the city. The dry weather flows are relatively low due to the small proportion of households connected to the system. The average annual rainfall is about 1200 mm, which falls during the rainy season (June–September) and accounts for about 70% of the annual rainfall with the highest peak in August and a smaller peak in April.

As a result of the deficiencies of the drainage system combined with the increasing urbanisation and loss of trees and other vegetation, the city suffers from serious regular flooding during the rainy season, when a large part of the runoff is stored in the low-lying parts of the city, blocking the traffic. In addition, the area has serious flood problems caused by the Bantyiketu river.

In the stormwater system almost all curb openings are clogged with sediments and solid waste, and some are even covered by new asphalt pavements. Many pipes show significant depositions and there is a significant wastewater inflow to the stormwater drainage system in the housing areas. As a result, the river suffers from severe environmental degradation, mainly caused by polluted flow from surface drains and ditches, which receives the septage and sullage from the city. The rain washes much of the surface pollution into the streams during wet season, but in the dry season the flow in the streams consists mostly of uncollected wastewater.

The solutions for the flood protection contain measures of river basin management, such as reforestation measures and technical solutions, such as attenuation ponds and a study was undertaken to consider various strategies for renovation of the drainage system. One improvement option included the construction of a more extensive network of drainage channels to convey excess run-off towards the river drainage system. To account for the increased runoff from the urban area, the following large scale structural interventions were considered as flood mitigation strategies:

- Increased capacity of the river channels in the Bantyiketu river system.
- Construction of six attenuation ponds with a total storage capacity of 124,400 m^3.
- A large storage reservoir (storage capacity of 115,000 m^3) with an overflow weir.

The present state of the stormwater drainage system was analysed by applying a model of the central part of the city in combination with a catchment area model of the Bantyiketu river. The complete catchment area of the river was considered and the area divided into the urban-runoff basin and the remaining part of the

Bantyiketu river basin. In order to evaluate the existing capacity, time series of rainfall data from 1995 to 1996 were used.

Hydraulic overload of the drainage system and the Bantyiketu river basin were of primary interest. For these two systems, to assess the effect of overflow structures on receiving waters, a computer model was developed using the simulation software SMUSI (Mehler *et al.* 1998), which is a physically-based hydrological deterministic rainfall and pollution load model.

Once the model of the existing drainage system had been developed and calibrated, the various stormwater management and flood mitigation strategies were evaluated.

As mentioned above, one option involved the simulation of the runoff from the catchment under the current situation and comparison with the runoff from a reforested catchment. The degree of effectiveness of reforestation was examined through a rainfall-runoff calculation for natural areas based on the Soil-Conservation-Service procedure (Soil Conservation Service 1972). Figure 8.6 shows an example of the runoff hydrographs from the completely reforested catchment compared with the catchment under the present state conditions (at the time of the study).

The simulation results clearly showed that the capacity of the river system was not adequate to convey large discharges of runoff after a severe precipitation event. Further, the model enabled various proposed structural measures to simulated. The project costs for these structural measures on the Bantyiketu river system amount to approximately US$20 million, although it was considered questionable whether such measures can be financed in Addis Ababa.

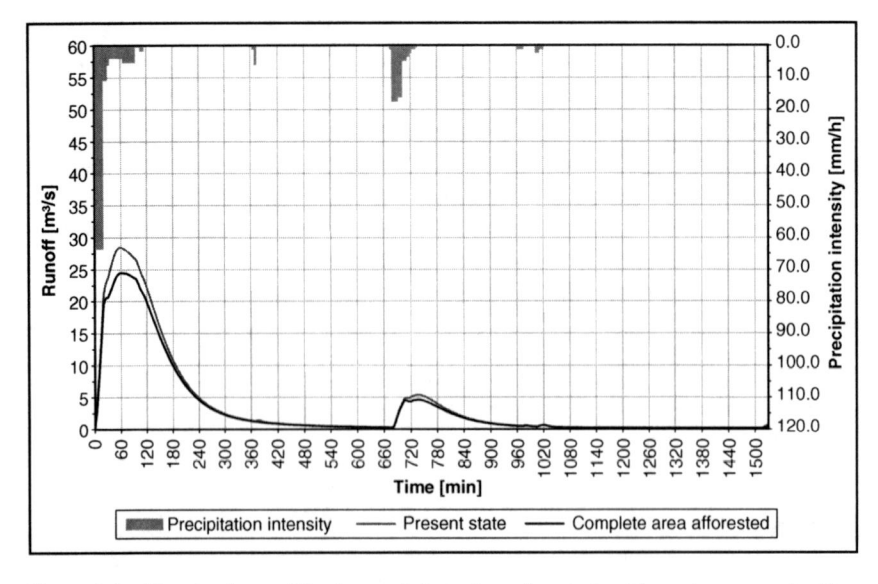

Figure 8.6 Simulated runoff hydrograph from the reforested catchment compared with the runoff hydrograph under existing conditions.

8.4.2 Application of flood simulation results

Results from the simulations may either be presented in the form of flood maps which indicate flood areas (see Figure 8.7) or may be linked via a GIS together to a digital terrain model (see Figure 8.8) and a land use map for detailed analyses of flood extent and flood damages. The model results must be geo-referenced and

Figure 8.7 Example of a 2-dimensional flood map from Thessaloniki, Greece (Antonaropoulos *et al.* 2003). The flooding points are highlighted with a circle.

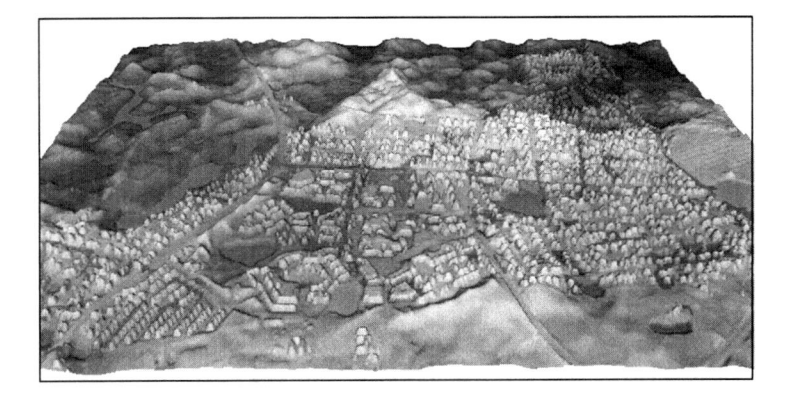

Figure 8.8 Example of a 3-dimensional flood map from Ballerup in Denmark. The light grey colour shows the ground and houses in the digital terrain model, whereas the dark grey colour indicates flooded areas.

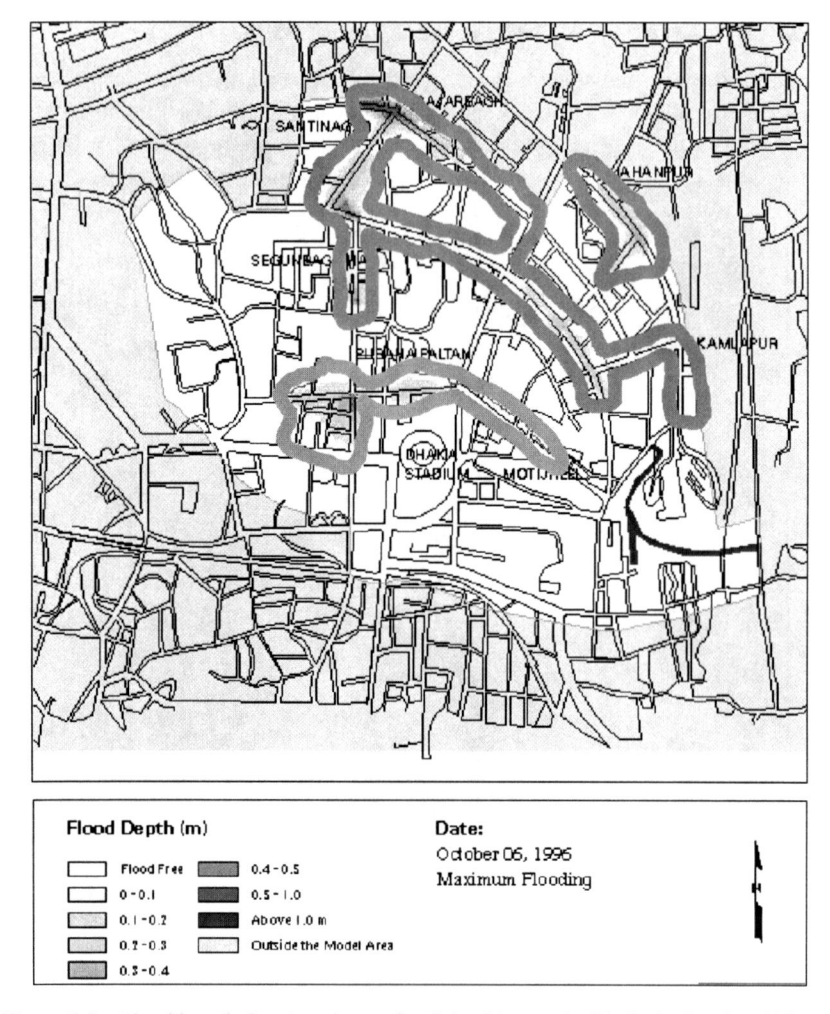

Figure 8.9 Flood inundation (maximum flood depth) map for Dhaka in October 1996. The bold lines show the flooded area boundaries as recorded by the city drainage authority. The white area is the model coverage.

related through a co-ordinate system linked to the digital elevation model (DEM). These results can then be presented in the GIS as flood inundation maps, based on the water levels computed by the urban drainage model. Flood inundation maps provide an effective media for visualising flooding. The water levels on the surface mainly cause flooding on the streets and the adjoining areas. The output of the simulation, in form of simulated water levels along the street system, can then be transferred to a GIS. Using interpolation routines, continuous three-dimensional water surfaces can be constructed based on the simulated street water level from the model and the DEM. The DEM elevations are subsequently subtracted from the water level surface delineating inundated areas by flood extent and flood depth. Water level results from the simulation are available along the streets as shown in Figure 8.9, which shows a flood inundation map during a major storm in Dhaka, Bangladesh in 1996.

8.4.2.1 Using the results

The model may now be applied for its specific purpose (e.g. management decisions or structural design). Presentation of the results should be accompanied by an assessment of its uncertainties. These may be quantified by a sensitivity analysis of its parameters and through a statistical analysis of errors in the validation period.

Case study: Urban flooding in Dhaka

In Bangladesh, the Dhaka Metropolitan area has experienced water logging for many years. Even a little rain can cause serious flooding problems in some parts of the city. The water depth in some of the flooded areas can be as much as 0.5–0.7 m, which creates major infrastructure problems for the city. The city of Dhaka is protected from river flooding by an encircling embankment. Most of the time during the monsoon season, the water level in the river remains higher than the water level inside the city area. This means that the city drainage depends very much on the water levels in the peripheral river systems. Consequently, conventional gravity-flow drains are not always sufficient to alleviate the flooding problems. To facilitate drainage, there are plans to install pumps at some of the outlets to the rivers. Furthermore, major reconstruction work has also been proposed and by using modelling this can now be evaluated before its implementation.

In order to analyse the urban flooding, the pipe flow system and its interaction with the flow on the ground surface was modelled. An exchange of water from the pipes to the streets (and vice versa) takes place, depending on the local hydraulic conditions in the catchment. To define the water routing on the surface, a DEM was generated in ArcView based on geographical information from the catchment. The output from the model simulations of the networks and the catchment consists of water levels and flows in the pipes and on the surface. A hydrodynamic model simulating the flow in the pipes and the surface flow simultaneously was used for this scenario. To validate the model, flood maps from the simulations were compared with the best available information on flooded areas from the October 1996 rains. An example of a flood inundation map can be seen in Figure 8.9 which shows good agreement between the observed flood extent and the model prediction.

As a part of the drainage improvement plan for Dhaka Metropolitan city, the Dhaka water supply and sewerage authority decided to rehabilitate the natural channel section of the Segunbagicha Khal (an urban drainage channel) by replacing it with a concrete conduit which has a total length of 2.1 km. A hydrodynamic computer model was used to evaluate the impact of the new conduit and to qualify the hydraulic processes involved in urban flooding in Dhaka (including evaporation and infiltration of groundwater).

The model was used to simulate proposed alleviation scenarios, such as construction of pump stations and application of real-time control for pump and sluice gate operation, and these simulations were used to evaluate the potential impacts of implementation. Some of the findings from the modelling of urban flooding in Dhaka include:

- The extent and the duration of flooding can be accurately reproduced using modelling tools.
- The conduit alone will not solve flooding problems in the catchment. A change in management strategies for the pumps is also needed to reduce the flooding in order to achieve the full benefit from the new conduit.
- Physical processes such as evaporation from catchment surfaces and infiltration of groundwater to the storm drains are insignificant during the flooding period.

8.5 INTEGRATED MODELLING OF THE URBAN DRAINAGE SYSTEM

In most parts of the world, planning and management of the sewerage system, the wastewater treatment plant and the receiving waters have been fragmented. Furthermore, there is a general lack of regulatory framework and technology available for the integrated planning and management. An exception to this rule is the Urban Pollution Management (UPM) procedure (Foundation for Water Research 1998) in the UK, which is a highly recommended reading. The challenge of today is to move the legislation from an individual consideration of each sub-system performance to an integrated management of the urban wastewater system. A step further in this direction is the European Water Framework Directive. The logic behind integrated management has been identified for some time, but it was the advent of computer models that was the more important single factor in the development of integrated analyses.

Integrated modelling in urban drainage is a term describing the modelling of the interaction between urban drainage systems, the receiving waters[3] and wastewater treatment plants. Integrated modelling is a challenge – not only due to the complexity of the model problem, but due to the different modelling approaches for each sub-system that have developed during the course of modelling history. Hydrodynamic data such as flow and water level can be directly transferred between the sub-systems, but water quality aspects are approached very differently by types of models, for example between rivers and the wastewater treatment plant. This can

---

[3] The term 'receiving waters' is used here to describe all natural water bodies (i.e. groundwater, streams, rivers, lakes and the sea).

cause significant problems in the interface between these systems in an integrated modelling exercise.

The quality of the output from an integrated model of an urban wastewater system depends strongly on the process description in sewer model, which today is the weakest link in integrated modelling. Despite these uncertainties, integrated modelling is still very beneficial for analyses, for example, in the assessment of the impact on water quality from sewer overflows.

Many urban areas have drainage systems interacting with the receiving water. However, many cities in developing countries do not have wastewater treatment plants. Hence, in order to simulate existing systems, the integrated modelling will often require a description of the interaction between the urban drainage system directly with the receiving water and not via a treatment process.

REFERENCES

Antonaropoulos, P., Karavokiris, G., Tselentis, J., Tomicic, B. and Pretner, A. (2003) Master plan of Thessaloniki's sewage network. *International Conference on Application of Integrated Modelling*, Integrated Modelling User Group (IMUG) Tilburg, The Netherlands. 23–25 April.

EU (2000). Water Framework Directive (2000/607EC), European Union.

Foundation for Water Research (1998) *Urban Pollution Management (UPM) Manual – A Planning Guide for the Management of Urban Wastewater Discharges during Wet Weather*, 2nd edn. Foundation for Water Research Report FR/CL 0009, November.

Mark, O., Weesakul, S., Apirumanekul, C., Boonya Aroonnet, S. and Djordjević, S. (2004) Potential and limitations of 1D modelling of urban flooding. *Journal of Hydrology* 299(3–4), 284–299.

Mehler, R., Leichtfuß, A. and Ostrowski, M. (1998) *SMUSI*, Institut für Wasserbau und Wasserwirtschaft, Fachgebiet Ingenieurhydrologie und Wasserbewirtschaftung, TU Darmstadt.

Muschalla, D. and Ostrowski, M. (2000) Urban storm water drainage system in the central part of Addis Ababa, Ethiopia – present state and proposals for the improvements. *9th International Conference on Urban Storm Drainage*, Portland, Oregon, USA. 8–13 September.

Johansen, N. B., Linde-Jensen, J. J. and Harremoës, P. (1984) Computing combined system overflow based on historical rain series. Chalmers University of Technology, *3rd International Conference on Urban Storm Drainage*, Göteborg, Sweden.

Rauch, W., Bertrand-Krajewski, J.-L., Krebs, P., Mark, O., Schilling, W., Schutze, M. and Vanrolleghem, P. A. (2002) Mathematical modeling of integrated urban drainage systems. *Water Science Technology* 45(3), 81–94.

Refsgaard, J. C. (1995) *Set-up, Calibration and Validation of Hydrological Models*. Danish Hydraulic Institute, Denmark.

Schütze, M., Campisano, A., Colas, H., Schilling, W. and Vanrolleghem, P. (2004) Real-time control of urban wastewater systems – where do we stand today? *Journal of Hydrology* 299(3–4), 335–348.

Soil Conservation Service (1972). *SCS National Engineering Handbook, Section 4, Hydrology*, US Department of Agriculture.

9
Operational performance and maintenance

Many infrastructure improvement projects direct considerable attention to design and construction, but few pay sufficient attention to the long-term operation and maintenance requirements. Deficiencies in operation and maintenance (O&M) inevitably mean that drainage systems do not operate in the way for which they were designed and consequently the expected level of benefit is not achieved. The deterioration of infrastructure assets due to a lack of O&M represents an enormous financial loss resulting in reduced asset life and premature replacement. Neglecting maintenance also results in increased cost of operating facilities, and a waste of related natural and financial resources (UNCHS 1993).

This chapter focuses on the issues related to O&M, and the need to focus on improved operational strategies in order to achieve performance objectives. Particular attention is directed towards the importance of the need to manage solid wastes effectively, as the build-up of solid waste and sediment in the drains and other problems related to solid waste is the prime cause of operational problems in drainage systems.

9.1 OPERATIONAL SUSTAINABILITY AND PERFORMANCE EVALUATION

Operational sustainability requires that drainage systems continue to offer an adequate level of service in relation to the design objectives over a prolonged period of

© 2005 IWA Publishing. *Urban Stormwater Management in Developing Countries* by Jonathan Parkinson and Ole Mark. ISBN: 1843390574. Published by IWA Publishing, London, UK.

time. Operational sustainability does not mean that the system does not require any attention to O&M, but it does mean that it can be operated over a period of time using the available resources without the need for major corrective intervention.

Operational sustainability is dependent on a wide number of factors – but a critical issue relates to the need for sufficient financial resources. Many local governments are reluctant to spend their limited resources on O&M, particularly due to a lack of cost recovery for urban drainage services (see Chapter 12). This factor, combined with a prevailing attitude of neglect towards the upkeep of assets, generally means that insufficient attention is paid towards O&M. As a result, there is generally a lack in systematic maintenance procedures in developing countries and the operational performance of stormwater facilities almost invariably deteriorates over a period of time.

Consequently, the majority of stormwater infrastructure does not function as effectively as it should and there is often a large gap between the theoretical design capacity of the drainage system and how it actually operates in practice. In addition, there is generally no or little attempt to monitor the performance of the service and most maintenance interventions are in response to a dramatic failure in system performance or continued public complaints about the deficiency of the system.

9.1.1 Measurements and indicators of operational performance

Measures or indicators of operational performance are necessary to monitor the performance of the drainage system, which can help to guide future policies (see Chapter 5) as well as having a direct relevance to current maintenance requirements. In general, these refer to both operational performance related to flooding and pollution impacts in receiving waters.

In relation to flood protection, there are three main questions to be considered that are particularly important from the perspective of residents who live in areas at risk from flooding (Kolsky 1998):

- How often does it flood?
- Where will it flood?
- What happens when it floods?

There are a number of different approaches that may be used to assess and monitor the performance of an urban drainage system. These are described below and summarised in Table 9.1.

9.1.1.1 Performance measurements

One way of monitoring performance is to directly measure parameters such as discharge, depth of flooding, pollutant loads, etc. but although these measurements may be used to quantify the problem, the relevant data are generally difficult to obtain. Also, it is not always obvious how to translate this information into operational strategies to improve system performance without extensive data analysis.

Table 9.1 Measurements and indicators of operation performance (after Kolsky 1998).

Type of measurement	Examples	Advantages	Disadvantages
Performance measurements	• Frequency of surcharge, flooding or overflow • Depth, area and duration of flooding • Volume and concentration of discharge	• Provide direct measurements of system performance	• Difficult to measure • Link to remedial action is not well defined
Performance indicators	• Solids levels • Inlet blockage • Hydraulic capacity	• Easier to quantify than process indicators • Clearer link to action than performance measurements	• Relation to outcome may not be clear • May measure symptom but not problem
Process indicators	• Frequency of cleaning • Staff time allocated to operations	• Relatively easy and practical to measure routinely • Clearly defined course of action	• Relation to outcome may not be clear

9.1.1.2 *Performance indicators*

As alternatives to direct measurements of system performance, performance and process indicators may be used to assess operational performance. Performance indicators are easier to measure than the performance measurements described above and there is also a more clearly defined course of action in relation to remedial measures that need to be undertaken.

9.1.1.3 *Process indicators*

Process indicators relate to the O&M activities that are adopted by the drainage system operator to maintain the operational performance of the system. As well as being easier to measure, these can be more useful to apply in practice in terms of routine application than performance indicators. However, as the relationship between process indicators and system performance is not always well defined, there is a risk that changes in the process indicators do not reflect the desired changes in the performance of the system (Kolsky and Butler 2002).

Measuring data is only one part of a strategy for performance evaluation. The other part is to know what to do with the data once these have been collected. In developing countries, the majority of urban authorities responsible for stormwater management lack any systematic way of monitoring and evaluating the quality of their

> **Box 9.1 Performance indicators for wastewater services**
>
> A methodology developed by the International Water Association's (IWA) O&M Specialist Group uses performance indicators to assess the operational performance of a wastewater system including urban drainage – either to see how the performance is changing over a period of time, or to compare with the performance of another system (benchmarking). This approach enables utility managers to make decisions based upon a standardised procedure. The methodology uses a software programme designed to simplify an otherwise complex procedure, which can be used to guide the process of selecting and implementing a system of indicators for the utility. The key issue to the applicability of the methodology in developing countries is the availability of reliable data for the input variables and application of the software. However, the software has been designed in a way in which a few or many of the parameters may be utilised.
>
> *Source*: Matos *et al.* (2003)

services. However, there are standardised procedures and methodologies to measure the performance of urban wastewater systems. For example, Box 9.1 describes a management tool that has been developed specifically for urban wastewater systems, including the urban drainage system. These methodologies have potential application in developing countries as a means to evaluate and monitor system performance.

9.2 OPERATION AND MAINTENANCE (O&M) STRATEGIES

All drainage systems, irrespective of their design and construction, require attention to O&M. But some require more attention than others. This will depend on the quantity and types of solid waste in the drainage system, combined with climatic factors that affect the duration of the wet season and the accumulation of sediment.

O&M strategies for urban drainage systems in developing countries are very different from those in developed countries due to cheaper labour costs. The strategy adopted will also depend on the type of drainage system itself. The majority of urban drainage systems designed as gravity flow systems require little in the way of regular operational activities compared with other urban infrastructure and services. However, those that rely on pumping or operation of sluice gates will require more attention to the operation of the equipment in order to ensure that the drainage system functions.

Also, in general, separate surface drains are easier to maintain and clean than combined systems, but it is generally cheaper and easier to implement a maintenance programme for single combined systems. Making a distinction between O&M activities can prove to be difficult, but in general *operation* deals with the running of a service on a daily basis, whereas *maintenance* deals with the less frequent activities that are necessary to keep the system in proper working condition.

By the nature of the drainage system, the need for O&M tends to occur during the wet season – which is not the best time to carry out these activities, except where the system needs emergency repairs. Although municipal agencies often make an attempt to clean and improve the operation of the system prior to the onset of the wet season, this tends to be only a partial response to the scale of the problem.

Due to the fact that maintenance strategies are not required for everyday operations, they are often not planned or implemented effectively. The practice most commonly adopted by municipal agencies responsible for management of drainage infrastructure is to clean the drains before the rainy season. It may therefore be beneficial to adopt maintenance strategies that remove the waste from the drains more frequently throughout the year rather than a major operation once per year, which requires significant resources, in both human and financial terms.

The different types of maintenance procedures for urban drainage systems are described below:

9.2.1 Routine maintenance

Routine maintenance activities are required on a regular basis throughout the year, but are often most frequent before and during the rainy season. In general, routine maintenance activities focus on preventative activities, which include unblocking of drains, clearing debris and vegetation from screens, culverts and tributary drains. Routine maintenance follows a defined schedule of activities and should lead to improved operational performance, reduced need for other forms of maintenance and extended lifespan of the infrastructure. In addition, routine maintenance can spread costs over a period of time.

9.2.2 Periodic maintenance and rehabilitation

Periodic maintenance is orientated towards rectification and rehabilitation of structural deficiencies. The scale of these activities may vary considerably from small-scale repairs on drains to larger-scale rehabilitation of culverts, etc. Rehabilitation work may be extensive enough to require engineering drawings and contracting in the same way as new construction. The timing of this type of work is difficult to define and will depend on a system for monitoring to identify when rehabilitation is required. Proactive maintenance involving inspection and repair should be focused on those parts of a drainage system that exhibit a tendency towards early failure, but this requires the development of a systematic database to monitor these trends.

9.2.3 Emergency (reactive) maintenance

Maintenance activities responding to emergency situations tend to be highly expensive and should be avoided if at all possible, except where valuable assets and human resources are put at risk by system failure. Therefore, they should only be in response to emergency situations calling for immediate action. The activities associated with this type of work include collapsed drains and obstruction of the flow by major obstacles such as fallen trees. Emergency maintenance should be minimised by placing greater emphasis on planned routine and periodic maintenance activities.

9.3 SOLID WASTE AND IMPACTS ON OPERATIONAL PERFORMANCE

As mentioned previously, the main factor that affects the operational performance of urban drainage systems is solid waste. An example of an urban drainage channel being used as a site for solid waste dumping can be seen in Figure 9.1. Although solid waste is widely acknowledged to be a major problem, the majority of municipal agencies have little data available on the nature and quantity of litter in stormwater drainage systems.

There are many sources of solid waste, and the types and constituents are summarised in Table 9.2. According to Hall (1996), the main factors influencing the quantity of solid waste, which can enter the drainage systems include:

- The failure of street sweeping services, lack of provision of bins and inadequate litter collection practices.
- Illicit dumping of solid wastes and the failure by authorities to enforce penalties to deter offenders.
- Excessive packaging and the irresponsible behaviour of residents to dispose of wastes.

The problem of refuse in stormwater drainage systems is often found to be at its worst where modern technologies, such as the plastics industry, have been introduced before the development of a strong environmental lobby or a policy for the waste management (Armitage and Rooseboom 2000). The problems became so severe

Figure 9.1 Excessive amounts of litter in a drainage channel in Dhaka, Bangladesh. (Photo: Birgitte Helwigh.)

146   Urban stormwater management in developing countries

Table 9.2   Types and constituents of solid waste (Armitage and Rooseboom 2000).

Type	Constituents
Plastics	Shopping bags, wrapping, containers, bottles, crates, polystyrene packing, etc.
Paper	Wrappers, newspapers, advertising flyers, bus tickets, food and drink containers, cardboard.
Metals	Foil, cans, bottle tops, car number plates.
Glass	Bottles, broken pieces.
Vegetation	Branches, leaves, fruit and vegetable waste.
Animals	Animal carcasses.
Construction material	Shutters, planks, timber props, broken bricks, lumps of concrete.
Miscellaneous	Old clothing, shoes, rags, sponges, balls, pens and pencils, oil filters, tyres.

in Dhaka, that 30% of the city's sewerage system became clogged and an 8-m layer of polythene on the bed of the Buriganga river that runs through the city was formed. As a result, in January 2001, the governmental authorities made the decision to ban the production, marketing and use of wafer-thin polythene bags as a result of the fact that at least five and half million of these plastic bags were being discarded every day (Roy 2002).

Other types of solid waste that cause significant problems in sewers are sediment, sand and gravel. These are generated by soil erosion from building sites and unsurfaced areas, and can have a significant implication on the performance of drainage systems and the pollutant loads discharges into receiving waters. Kolsky and Butler (2000) report measurements of the depth and size distribution of solids in an open drain in Indore, Madhya Pradesh in India. As illustrated in Figure 9.2, the size of the particles was observed to be a magnitude larger than those reported in drains in Europe.

9.3.1   Impacts of solid waste on operational performance

Uncollected solid wastes enter surface drains and sewers causing blockages and reduced flow capacity. Sediments in sewers also reduce the hydraulic capacity of sewers due to decreased cross-sectional area and increased hydraulic roughness. Flooding is more likely to occur if solid waste accumulates and blocks the drains and their inlets. Blocked drains may also create insect breeding sites and encourage disease transmission (see Chapter 2).

Excessive sediment deposits or blocking of inlets to the drainage system (see Figure 9.3) will inevitably result in operational problems resulting in premature surface flooding and overland flow of storm runoff. In addition, although pollution problems may not be perceived by residents to be as much of a problem as flooding, the quantity of pollution loads discharged from CSOs is affected significantly by the concentration of sediment that is washed out of the drainage system.

Operational performance and maintenance   147

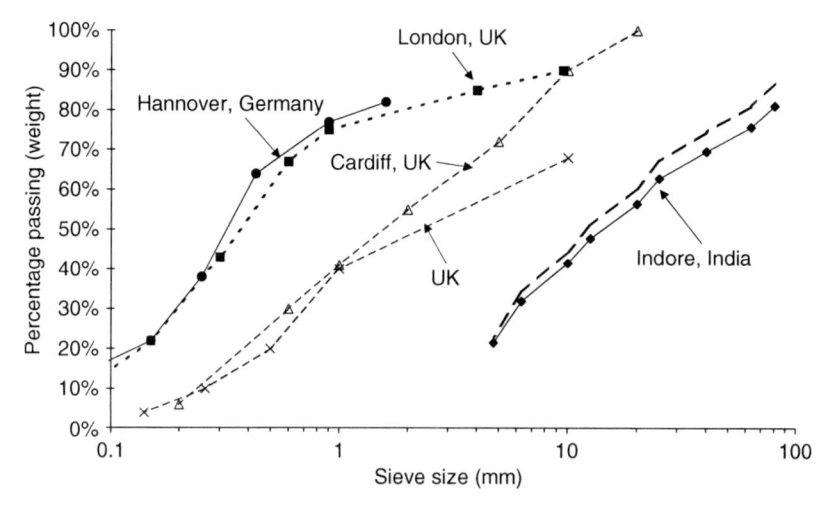

Figure 9.2   Comparison of solids sizes in Indore, India with samples from the Europe. (Adapted from Kolsky and Butler 2000.)

Figure 9.3   Blocked inlet to the stormwater drainage system. (Photo: Birgitte Helwigh.)

Solid waste and sediment also have significant impacts on the operation performance of retention and detention tanks. Piel *et al.* (1999) obtained results on the performance of various storage ponds in France. More than 200 different installations were evaluated accounting for storage of approximately 200,000 m^3. The research concluded that the performance of these facilities tends to deteriorate with time, especially where ponds are covered. In the majority of situations this is caused by poor maintenance due to difficult access, blocked openings or because the installations are forgotten after multiple changes of owners or site managers. The experiences from Brazil described in Box 9.2 suggest that the situation may be considerably worse in cities undergoing rapid development.

Box 9.2 Experiences in operation of retention basins from Belo Horizonte, Brazil

Experiences from Brazil highlight the problems related to the operational performance and maintenance of retention basins caused by rapid urbanisation combined with the lack of regulatory measures or methods for control of erosion, dumping of solid waste and illicit discharges of wastewater into surface water drainage systems. A typical example is the Santa Lucia retention basin in Belo Horizonte (see Figure 9.4), which has suffered from a significant reduction of storage capacity, degradation of water quality and aesthetic value of the pond due to excessive sedimentation.

The Santa Lucia retention basin on the Leitão creek was originally designed and constructed in order to alleviate problems from another retention basin (Acaba Mundo), which had a design capacity of 40,000 m^3, but suffered from such significant sedimentation that the municipal authorities decided to fill it in and create a park. However, the Santa Lucia retention basin, which is 7.5 times larger

Figure 9.4 Two views of the Santa Lucia retention basin, Brazil. (Photos: Nilo Nascimento.)

than the one that existed at Acaba Mundo, suffered similar problems and in 1976, despite an initiative to construct small sedimentation dams to reduce the solids loading, the basin was more than 65% full of sediments and solid waste. The first attempt to solve this problem was to construct a large drainage pipe from the bottom of the basin to enable flow of water (and sediments) to pass through the basin and flow out the other side. However, this inevitably meant that the benefits of the storage capacity from a flood control perspective were lost and the first extreme event caused massive downstream flooding and considerable disruption and damage.

As a result of political pressure, it was decided to investigate the possibility of reinstating the storage reservoir into the drainage system, but with a smaller volume (120,000 m^3). The need for the retention basin as a flood control device has heightened awareness and changed perception, not only of the public authority but also of the inhabitants of Leitão river basin, who have a clear interest in preserving Santa Lucia flood control functions.

Source: Nascimento *et al.* (2000)

9.4 CONTROL OF SOLID WASTE PROBLEMS

By far the most effective strategy for controlling solid waste is to stop it entering the drainage system in the first place, and the most effective way to do this is to ensure that there is an efficient solid waste collection system operating in the city. The provision of solid waste management services is outside the scope of this book and the reader is advised to consult other publications on the subject (see Appendix A1). However, common to all effective solid waste management strategies is the need to promote awareness about the problems. For example, Box 9.3 describes the importance of educating municipal workers and communities on the links between drainage systems, poor sanitation and transmission of diseases.

There are a number of activities related to the control of solid waste that are directly relevant to stormwater management. According to Butler and Clark (1995), sediment can be controlled by one or a combination of the following strategies:

(1) Surface cleaning and sweeping.
(2) Collection and removal in gullies or other traps at the entry to the drainage system.
(3) Cleaning the sewers where material has deposited.
(4) Removal at the outfall or treatment works.

In theory the most cost-effective place to collect and remove the sediment would be at the treatment works, but this is dependent on the design of the drainage system so that sediment does not accumulate in the pipes and drains. Therefore, this strategy does not work too effectively in practice and other strategies for the control of sediment and solid waste need to be considered.

> **Box 9.3   Training manual for drainage maintenance workers**
>
> This manual was designed for workers in order to educate them about the importance of improved sanitation conditions in reducing disease transmission, not only of mosquito-borne diseases such as dengue, malaria, yellow fever and West Nile virus, but of others related to bacterial pathogens and carried by rats and flies. It contains useful illustrations and diagrams that enhance its use as an educational tool – not only for the workers who clean and repair secondary drains for ministries and local governments, but also for the general public by making them aware of the role drains play in public health and safety. It is a good example of how social communication can be used to raise community awareness of domestic-sanitation issues and their effect on public health.
>
> *Source*: PAHO 2000

9.4.1   Gully pots and sediment traps

Various types of inlet control and sediment/solids traps are available, which contribute towards reducing the ingress of solids into the drainage system. Butler and Clark (1995) highlight the importance of the gully pot as an effective sediment trapping mechanism. Provided a routine cleaning operation is implemented, these devices have the greatest potential for overall cost-effectiveness of a sediment control strategy in a catchment.

9.4.2   Street sweeping

The objective of street sweeping, in addition to improving the quality of the urban environment, is to reduce the amount of solid waste and sediment that enters the drainage system. This can prove to be a cost-effective strategy for reducing problems in the drains themselves and can reduce the costs associated with drain cleaning. Street sweeping is important to reduce the potential for sedimentation and blockages in the drainage system, and may improve the aesthetics of urban areas by removing debris and litter from streets. However, this may have a limited effect in relation to pollution control due to the fact that many of the pollutants are attached to finer sediments, which are not collected unless the street sweeping is very effective.

Therefore even the most effective solid waste management strategy and street sweeping will not stop all solid waste, particularly sediment, from entering the drainage system. Although street sweeping is important for the control of large solid waste entering the drainage system, its effectiveness is limited as a sediment control strategy. Increasing the frequency of street sweeping practices beyond what is required to meet aesthetic objectives is not expected to yield substantial incremental benefits in relation to the receiving water quality improvements (Walker and Wong 1999).

> **Box 9.4 Drain cleaning and solid waste management in Lahore, Pakistan**
>
> The drainage system in and around the old, walled city of Lahore was constructed by the British during the colonial era. The main sewers are of brick construction with a large, egg-shaped cross-section. The system is a combined system and, over the years, became choked by deposited sediment and solid wastes, which blocked the flow of runoff. As a result, the drainage system became ineffective and, during the monsoon season, floodwaters caused structural damage to buildings, disruption to the traffic flow and an increase in the prevalence of waterborne diseases transmitted by polluted water.
>
> A project was initiated in 1992 to alleviate these problems by reducing the accumulation of the solids in the drainage system combined with an initiative to improve solid waste management in order to reduce the potential for reoccurrence of the problems in the future. The project was funded by UK Government's Department for International Development (DFID) and implemented by the Water and Sewerage Authority (WASA) with technical assistance from the Danish consultancy Carlbro in partnership with a local non-governmental organisation called Youth Commission for Human Rights (YCHR).
>
> Although the cleaning of the sewers themselves posed a challenge from a technical perspective, there were also many difficulties presented by the fact that it was impossible to undertake the cleaning activities during the daytime due to the amount of activity on the streets. The only time it was possible to carry out the work was during the night, but the disturbance and disruption during the cleaning of sewers and open drains (nullahs) would inevitably cause complaints amongst the local residents.
>
> The role of the YCHR was critical in liaising with the community about the project and communicating the reasons why the remedial work would be undertaken during the night. In addition to this, YCHR implemented an awareness-raising campaign to promote the importance of improved solid waste management in order to encourage residents to understand the linkages between poor solid waste management and blocked drains. In addition, the awareness-raising campaign was complemented by the introduction of a new system for collection of domestic refuse on a house-to-house basis. This involved development of a handcart, which had the mobility to access the narrow, crowded streets in the old walled city area.
>
> *Source*: YCHR, Lahore, Pakistan

9.4.3 Drain cleaning

Drain cleaning is by far the most common form of managing problems related to solid waste in drains, particularly in developing countries. Although some form of drain cleaning will always be necessary, it is not always the most efficient or cost-effective approach, and should not be the only strategy adopted for management of solids.

Box 9.4 describes a drain cleaning and solid waste management project in Lahore, Pakistan, which emphasises the importance of raising awareness about the need to manage solid waste to prevent future problems related to drainage blockages. The effective implementation of this project is dependent on community awareness and participation in solid waste management.

In the majority of situations, drain cleaning is infrequent and only in response to emergency situations. However, some municipal agencies try to tackle the problem, although often under-resourced. Box 9.5 describes the drain cleaning practice of the agency SADCO in Hanoi, Vietnam and links these to the importance

---

Box 9.5   O&M of drainage infrastructure in Hanoi, Vietnam

Sewer cleaning is a major task for SADCO, the organisation responsible for operating and maintaining the 180 km of the combined drainage system in Hanoi. SADCO currently serves about 60% of the city. In the other areas, responsibility for service provision is taken on by local authorities or under self-provision by local residents.

In addition to the solids from the wastewater flows, it is estimated that approximately 30% of total solid waste generated from the city is disposed into the drainage system and the channels and lakes in the city. There are four O&M enterprises under the responsibility of SADCO – each one is responsible for two districts. In the large sewers, the cleaning process is carried out by machines with hydro- or pneumatic flow-jets, metallic strings and rollers and vacuum tankers, and this is carried out all year long at a rate of about 15 km per month.

However, the activity is most intensive in the season before summer rains and this is when most of the manual cleaning is done. The majority of the manual cleaning is in small channels and sewers, but due a lack of equipment and resources, some of the larger drains have to be cleaned manually (see Figure 9.5). Approximately 24,000 m^3 of sludge is collected from the drainage system annually, of which 70% sludge is collected mechanically and 30% manually. The sludge and solid wastes are collected and transported by storage tankers, and then from these locations the wastes are collected and taken to the landfill disposal site.

Figure 9.5   Municipal workers manually removing silt from a drainage system in Hanoi, Vietnam prior to the onset of the rain season. (Photo: Jonathan Parkinson.)

> SADCO pays the workers for the work carried out based on the volume of sludge and garbage collected, but the operating budget of SADCO is still heavily subsidised by the State. In Hanoi, as well as in some other cities, a 10% surcharge on the water supply bill has been introduced for wastewater services. The deficit between revenue generated from the surcharge and operating expenditure is still high – but at least the concept of paying for the service has been introduced as low tariffs leads to wasteful use and generates inadequate revenues for water supply and sewerage system for their O&M and upgrading.
>
> *Source*: Dr Nguyen Viet Anh, CEETIA, Hanoi, Vietnam

of developing effective systems for cost recovery through a process of improved management practices and institutional reforms.

9.4.4 Storm drain flushing

Sewerage system and storage tank flushing reduces the deposited accumulation of solids and is an important maintenance strategy to optimise performance, maintain structural integrity and reduce pollution of receiving waters during wet weather. Storm drain flushing usually takes place along sections of pipelines with grades that are too flat to maintain adequate velocity to keep particles in suspension. An example of storm flush as part of a regular maintenance programme of the drainage system in Tehran can be seen in Figure 9.6, which includes a regular flushing and manual cleansing of the drains and debris screens.

Figure 9.6 Flushing and manual garbage collection of the drainage system in Tehran, Iran. (Photos: Ole Mark.)

However, flushing uses considerable amounts of water and is therefore only possible where there is sufficient water that is readily available from nearby watercourses. In addition, it may be better to adopt a strategy of sediment removal as the practice of sewer flushing causes discharges of polluted water into the receiving watercourse during dry weather when there is low dilution and this may have significant impact on the reduced water quality (Pisano *et al.* 1998).

9.5 REFERENCES

Allison, R., Chiew, F. and McMahon, T. (1997) *Stormwater Gross Pollutants*. Cooperative Research Centre for Catchment Hydrology. Industry Report 97/11. Monash University, Australia.

Armitage, N. P. and Rooseboom, A. (2000). The removal of urban litter from stormwater conduits and streams. Paper 1 – the quantities involved and catchment litter management options. *Water SA*, **26**(2), 181–187.

Butler, D. and Clark, P. (1995) *Sediment Management in Urban Drainage Catchments*. CIRIA Report No. 134, Construction Industry Research Information Association, London.

Hall, M. (1996) *Litter Traps in the Stormwater Drainage System*. Unpublished MEng Paper. Swinburne University of Technology, Melbourne (cited by Armitage and Rooseboom 2000).

Kolsky, P. J. (1998) *Storm Drainage: An Engineering Guide to the Low-Cost Evaluation of System Performance*. Intermediate Technology Publications, London.

Kolsky, P. J. and Butler, D. (2000) Solids size distribution and transport capacity in an Indian Drain. *Urban Water* **2**(4), 357–362.

Kolsky, P. J. and Butler, D. (2002) Performance indicators for urban storm drainage in developing countries. *Urban Water* **4**(2), 137–144.

Matos, R., Cardoso, A., Ashley, R., Duarte, P. Molinari, A. and Schulz, A. (2003) *Performance Indicators for Wastewater Services*. IWA Manual of Best Practice. IWA Publishing, London.

Nascimento, N. O., Ellis, J. B., Baptista, M. B. and Deutsch, J.-C. (2000) Using detention basins: operational experience and lessons. *Urban Water* **1**(2), 113–124.

PAHO (2000) *Drains and You – A Manual for Drainage Maintenance Workers*. Pan-American Health Association (PAHO)/World Health Organization, Trinidad and Tobago.

Piel, C., Perez, I. and Maytraud, T. (1999) Three examples of temporary stormwater catchments in dense urban areas: a sustainable development approach. *Water Science and Technology* **39**(2), 25–32.

Pisano, W. C., Barsanti, J., Joyce, J. and Sorensen Jr, H. (1998) *Sewer and Tank Sediment Flushing: Case Studies*. EPA Report EPA/600/R-98/157. US Environmental Protection Agency, Cincinnati, Ohio, USA.

Roy, P. (2002) A mixed bag. *Down to Earth*. Special Report, February 2002. Centre for Science and Environment, New Delhi, India, pp. 22–23.

UNCHS (1993) *Maintenance of Infrastructure and its Financing and Cost Recovery*. United Nations Centre for Human Settlements, Nairobi, Kenya.

Walker, T. A. and Wong, T. H. F. (1999) *Effectiveness of Street Sweeping for Stormwater Pollution Control*. Cooperative Research Centre for Catchment Hydrology. Technical Report 99/8, Australia.

10
Non-structural flood mitigation strategies

All urban drainage systems will invariably reach the limit of their hydraulic capacity at some point during their operational lifespan. It is therefore important to consider the implications of drainage system overload and to acknowledge that, in many situations, flooding is inevitable in the urban environment. The scale and duration of flooding will vary enormously, and it is the role of the urban drainage engineer to consider the implications of system overload, to identify where flooding is most probable, and to consider what is the best approach to reduce these risks.

Sometimes the most appropriate solution is to improve the hydraulic capacity, either by maintenance and rehabilitation of the existing system or by investments in new infrastructure. However, in many situations, especially where the risk of flooding is high, non-structural flood mitigation strategies are important measures to reduce the risk and damage caused by flooding. The key feature of these strategies is the emphasis on *non-structural* measures – meaning that they do not involve physical intervention or are in addition to basic physical interventions. This chapter considers what can be done to avoid or mitigate the impacts caused by flooding using non-structural FMS and considers the range of actions and remedial measures that form part of these strategies.

© 2005 IWA Publishing. *Urban Stormwater Management in Developing Countries* by Jonathan Parkinson and Ole Mark. ISBN: 1843390574. Published by IWA Publishing, London, UK.

10.1 STAGES OF THE FLOOD MITIGATION CYCLE

In order to formulate flood management and control policies, a thorough understanding of flood problems is required. This can be achieved through data collection and analysis along with flood impact assessment. Bearing in mind that no drainage system can be designed to convey all runoff in all storm conditions, the questions that need to be considered during the design process include:

- Where is it likely to flood when the system is overloaded?
- What will be the implications of system overload?
- Who will be affected?
- How can flood damage be reduced?

As described in Chapters 1 and 2, the range of problems related to flooding are diverse and therefore it is necessary to consider strategies that mitigate the impacts or, if possible, to avoid the problems altogether. These can be divided into the following categories:

- Preventative actions that consist of precautions in order to avoid the impacts of flooding.
- Mitigation measures that reduce the extent and scale of flood impacts.
- Remedial actions that help those affected by flood to recover from the damage and other negative impacts caused by flooding.

These may include awareness raising and activities required to promote appropriate technologies for residents to take action to protect their houses from flood damage (as described in Section 10.3).

Non-structural flood mitigation strategies rely upon action and support from households and local organisations working collectively, and require the participation from the inhabitants of flood-risk areas (see Chapter 11). In addition, flood warnings need to be issued so that communities can prepare for the onset of a large flood event and for urban authorities to prepare for an emergency situation. These response strategies can minimise potential damage, but there will also be a need to develop appropriate strategies for flood recovery and rehabilitation for affected communities. The example described in Box 10.1 illustrates the importance of a combination of these approaches for a comprehensive strategy for flood mitigation which takes into account aspects of flood-risk assessment and risk management. The framework for the development of risk assessment is illustrated in Figure 10.1.

As described in Chapter 2, many communities in flood-risk areas have developed complex adjustments to floods, which include their own intuitive awareness and indigenous flood-warning strategies. The complexity of these should not be underestimated and flood mitigation plans should take these as the starting point for developing flood mitigation and response strategies. As described in Chapter 11, discussions between urban drainage planners and community members,

> **Box 10.1 Non-structural flood mitigation measures for Dhaka city**
>
> After the disastrous flood of 1988, caused by a lack of flood protection and the excessively high upstream river flows, the Dhaka Integrated Flood Protection Project (DIFPP) was initiated as a part of the National Flood Action Plan in Bangladesh. DIFPP focused primarily on structural measures for flood damage mitigation, but when the city was hit by another catastrophic flood in 1998, it became apparent that structural measures alone were not sufficient to guarantee flood protection for the city. As a result, a variety of non-structural flood control strategies were proposed as part of a long-term flood protection and mitigation plan for the city.
>
> *Land use planning and development control*
> Land use development controls were developed based on geographical, topographical, climatic and soil characteristics and delineation of land for residential, business, industrial and natural uses. The planning process also involved categorisation of land usage based on risk of inundation including the identification of the most appropriate measures to protect the lakes and natural depressions which serve as retention ponds for runoff and are important to prevent inundation in the city, from encroachment by developers.
>
> *Improved solid waste management*
> Indiscriminate disposal of solid waste combined with runoff from construction sites heavily loaded with sediments was identified to be a major factor, contributing to the poor performance of the existing drainage system. A campaign to increase public awareness and to promote active public participation to reduce the problems associated with solid waste management formed an important component of the integrated urban stormwater management plan requiring non-structural interventions.
>
> *Institutional co-ordination*
> Improved inter-agency co-ordination and enhanced communication between the various municipal agencies (Dhaka Water and Sewerage Agency, Bangladesh Water Development Board and Bangladesh Agricultural Development Corporation) with direct involvement or interests in stormwater management was emphasised as being necessary to avoid some of the causes of flooding.
>
> *Flood proofing and warning*
> Both the general public and experts expressed a similar view that flood proofing and improved flood warning system are effective means of minimising flood damage. Individual preparedness was identified as being the best way to minimise personal loss due to flooding. This reflected people's view of the inadequacy of existing measures offered to flood victims and their preference for self-reliance in combating flooding.
>
> *Source*: Faisal *et al.* (2000)

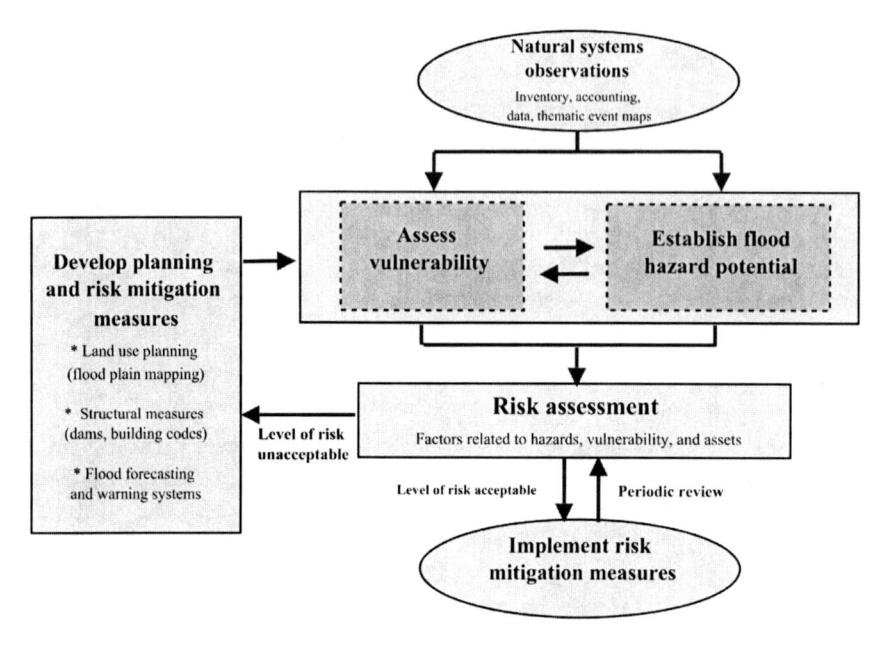

Figure 10.1 Framework for flood-risk assessment and risk management. (Adapted from ISDR 2001.)

assisted by community development workers experienced in participatory planning and social mediation, can also provide an opportunity to discuss community responses to flood events and their reaction to flood warnings.

Figure 10.2 illustrates the flood mitigation cycle and the five stages in the flood mitigation cycle, which form the basis for the development of a strategy for flood protection and response, are described below.

Review of past experiences

The first stage of the development of a flood mitigation strategy should be a review of past experience. In order to develop effective response strategies and to determine how responsibility for action should be delegated, this should include a consideration of the possible flood scenarios affecting the area and factors affecting the degree of danger in the floods experienced in the past.

However, it is important to consider that the local community may not have any experience with the extreme floods that occur very rarely (e.g. once in every 50–500 years). In addition, conditions may have changed significantly since the last flood event that is remembered by the local community. For instance, more recent urban developments may have increased the potential flood damage and flow conditions of floodwaters may also have been modified.

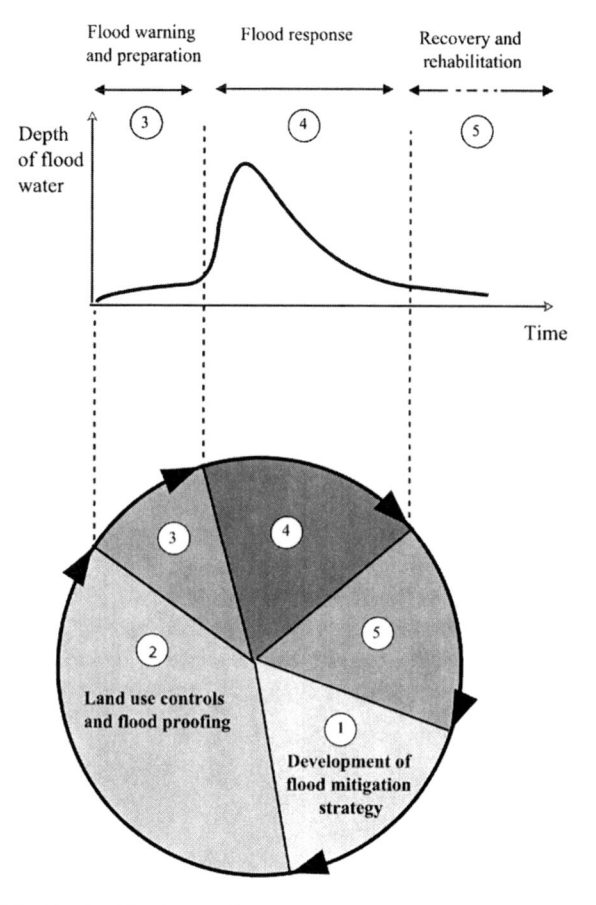

Figure 10.2 The flood mitigation cycle.

Land use controls and flood proofing

Land use controls aim to prevent construction of houses in low-lying areas prone to flooding or on steep slopes prone to landslides. Flood proofing involves various structural adaptations to buildings to reduce the impact of flooding. In addition to flood zoning, this may include implementation of other flood control measures, such as the designation of areas for attenuation of flood runoff storage. Flood mitigation strategies are implemented well in advance of the onset of floods and should form part of the flood mitigation strategy for the city as a whole.

Flood warning and preparation

Flood warning and preparation strategies need to be implemented as emergency flood responses to enable communities in flood prone areas to prepare for and respond to flood hazards. Preparedness planning is based on raising public awareness. These activities include monitoring and identification of potential flood hazards, and release of flood warnings whenever the flood hazard is considered to be significant and threatens a specific area. Preventative actions must be continuously monitored and updated to make sure to produce a timely and appropriate response to potential flood hazards.

It is important that the flood warnings are based on forecasts, which are as accurate as possible. The accuracy of these warnings should be taken into account and the warnings should be formulated to reflect the certainty of the forecast. For example, a slowly rising water level in a big river is relatively easy to forecast. On the other hand, flash floods in smaller catchments are more difficult to forecast and have a higher degree of uncertainty. It is therefore important to inform the public about the reliability of the forecast to maintain integrity of the flood-warning system.

Flood response

This stage in the flood mitigation cycle involves a range of activities that are required once the flood warnings have been issued. These will depend on the predicted severity of the event, but may include evacuation of people, closing schools and public offices, and preparation of emergency power and alternative water and food distribution.

Flood recovery and rehabilitation

Recovery and rehabilitation strategies after a flood event include a wide range of activities that are necessary to repair damage and return to normal life as soon as possible. Many of these will be undertaken at the household level, but there is an important role for urban authorities and non-governmental organisations (NGOs) to assist them.

10.2 FLOOD MITIGATION THROUGH LAND USE CONTROLS

As well as being a cost-effective means to control the generation of urban runoff, land use controls can be implemented to minimise exposure to the risk of flooding. These interventions are closely linked to land management and urban planning. One approach is to restrict construction on the flood plain, but as demand for land in cities is high, it may be necessary to consider alternative approaches for the control of developments on land which involve alternative uses for the land in order to ensure that informal settlements do not develop.

10.2.1 Flood zoning

A flood zone is an area of land that is prone to flooding and flood zoning is a planning tool, which is widely used as the basis for land use control. Zoning involves the designation of the type of activity that can be undertaken within the flood-prone area and can be used by urban authorities to help control the type of development or redevelopment allowed within their boundaries. Flood zoning is a planning tool which may be used to assist flood mitigation strategies by identifying the most appropriate types of land use and this approach can be used as the basis for development of legal measures for land development.

To assist in the development of the flood-zoning plans, it is advisable to develop a computer model to simulate the hydrological regime. As described in Chapter 8, a model can be used to assess various scenarios and the impact of different stormwater management strategies in terms of parameters of operational performance such as depth, extent and duration of flooding as a function of return period of the storm events. The main constraint to this is often the availability of a good data to calibrate the model as large-scale flooding generally happens infrequently. Therefore, although calibration of the models is often not possible for the extreme events, reasonably good estimates of model parameters can be made from experience about past flood events. The availability of accurate topographical data may be a limitation, which should not be ignored when the flood zones are planned.

It will be necessary to map the areas, which are identified to be susceptible to flooding and these maps subsequently form the basis of the flood-zoning maps. The flood zones are normally related to the probability of floods (e.g. separate zones will be drawn for a 1-, 5-, 10-, 20-, 50- and 100-year flood). The first stage of flood zoning should therefore be to identify the areas that are prone to flooding which are often well known to local residents. It is therefore important to consult residents in order to review past experiences and perceptions of flooding. Based on further consultation with inhabitants and those who use land in flood-risk areas, the flood zones can be developed. The main features of flood zones include information about how often it is likely to flood and the recommended land usage. This needs to be supported by legal mechanisms to prevent the development or redevelopment of illegal settlements, or other undesirable land uses. This may also be supported by financial and economic measures (e.g. land taxes), to discourage these developments or progressive taxes for recommended land use such as parks, sport fields or agriculture (Campana and Tucci 2001).

10.2.2 Resettlement

Where communities have already constructed houses on drainage pathways and floodplains, it may be necessary to relocate some families to areas where they are less exposed to flood risks. Experiences suggest that large-scale eviction and forced relocation may exacerbate social problems, which may be considered by those affected communities to be more severe to than the original flooding problems (Tucci 2002). Therefore, where resettlement is the only viable option, it is

> **Box 10.2 Disaster mitigation in landslide and flood prone areas of Bogotá, Colombia**
>
> As a response to the problems-facing communities living in areas at risk from landslides, mud-flow, falling rocks and flooding, the municipality of Bogotá took an initiative to protect the vulnerability more than 6000 households in high-risk areas. The first step was to identify the areas of risk and to develop a strategy for risk reduction based on mapping of the highest-risk areas. The risk zonification allowed for the development of a plan for structure of land use resulting in a control and protection of soils. It is necessary to have a good working technical knowledge of the threats and Geographical Information Systems (GIS) are an important tool in this respect to be able to assist in the development of appropriate intervention measures. Where possible, households were protected by structural interventions to reinforce slopes and strengthen houses. However, in many situations, the only option to protect the families was relocation and resettlement. The resettlement process was designed to guarantee the improvement of living conditions of families, including safety, public utility service availability and mobility among others. To ensure sustainability of the resettlement and to ensure that the risk areas were not taken over by families again, the legal market of development lands must be competitive with the illegal housing market. In this respect, housing offers were structured for highly social vulnerability sectors whose incomes were below the minimum legal wage. Of importance to the success and sustainability of the risk reduction interventions was the role of the municipality to formalise policy and official planning, and intervention procedures that were adopted by the project.
>
> *Source*: Together Foundation and UNCHS

necessary to ensure that resettlement plans are designed and implemented with the groups who are being resettled (especially with regard to the choice of the relocation site). In order to ensure that people want to stay in the areas of resettlement, it is also important to consider the provision of appropriate infrastructure and services in combination with opportunities to support their livelihoods (Santosa 2003).

Box 10.2 describes an example of a resettlement programme in Bogotá, Columbia. The success of the resettlement process can be seen to be dependent on strong levels of community organisation and by their involvement in the design, planning and implementation of the resettlement programme. In many situations the role of NGOs is important as mediators between affected communities and municipal authorities.

10.3 FLOOD PROOFING AND BUILDING CONTROLS

Land use and zoning policy cannot entirely eliminate the presence of hazards. As an alternative strategy to relocation, flood proofing provides structural protection

Non-structural flood mitigation strategies

Figure 10.3 Example of temporary flood proofing – house constructed on stilts to avoid flooding. (Photo: Birgitte Helwigh.)

Figure 10.4 Example of temporary flood proofing. (Photo: Birgitte Helwigh.)

to housing and buildings to reduce the damage caused by flooding and inundation. Houses may be constructed to reduce the flood risks to the individual household. This may involve the use of landfill to raise the foundations of the house or by building the house above the natural ground level using piles or stilts (see Figure 10.3).

Other flood proofing involves construction of floodwalls around houses to hold back floodwater (see Figure 10.4) or design of structures to withstand the effects of inundation. Structural adaptations to houses may involve raising plinth levels, paving courtyards, using landfill to raise the structure and building houses out of materials, that are resistant to flooding. This may involve sealing houses to prevent floodwaters from entering the building (dry proofing) or making the structure more resistant to flood damage (wet proofing).

Some of these measures are most effective when implemented at the community level. For example, a floodwall may be best introduced so that it protects more than an individual house, and it might be possible for the community to work collectively in its construction. However, it is important to remember that the construction of floodwalls or other structures preventing floodwater to enter a part of the flood plain may increase the flooding problems for communities downstream. The planning of these measures should therefore be done jointly taking all affected areas into account.

Application of building codes and construction specifications is an issue that requires a certain amount of flexibility, because codes may entail unaffordable investments and therefore can be restrictive, especially for low-income families – many of whom will be the ones who are most at risk from flooding and most vulnerable to its impacts. If these standards are too high, the cost of materials may mean that these strategies are not affordable by poorer communities. Also, poor communities in informal settlements, especially those who lack security of tenure, are just as likely as to face demolition or forced relocation as they are destructed by flood events. Therefore, residents are unlikely to want to invest in household improvements unless their land tenure is secured and will therefore continue to use very low-cost materials for construction.

Low-income communities, who are most vulnerable to floods, are especially versatile in the use of low-cost structural adaptations to protect their houses from flooding. In Indore, India, households and small enterprises were observed to make permanent and temporary adaptations in response to the risks of flooding and inundation. Residents may also purchase metal furniture that is more resistant to immersion and ensure that shelving and electric wiring are high up the walls, above expected water levels. In some very low-income areas, roofing was observed to be weighed down with rocks rather than attached to a house, so that the roofing can be removed quickly if the structure is in danger of being swept away (Stephens *et al.* 1996).

10.4 FLOOD RESPONSE STRATEGIES

As shown in Figure 10.5, a lack of adequate systems flood warning can result in considerable disruption – much of which could be alleviated if residents are informed in advance of the imminent flood threat. For this reason, *pre-disaster planning* is considered to be the key to effective response strategies. Public awareness of flood risks and emergency management procedures include all social groups during all phases of disaster planning and management.

Figure 10.5 Lack of flood warning catches residents by surprise and causes chaos on the streets of Bangkok. (Photo: Ebbe Worm.)

Strategies for flood-emergency response relate to action that needs to be taken both *immediately before and during a flood* in order to reduce the damage and impacts on communities. The foundations of a flood-emergency action include a well-designed mobilisation plan with an effective communications and public information strategy. The operation of flood control works is dependent on advance warning and inter-flood responses depend on effective systems to monitor and assess the flood situation.

As described in Box 10.3, one of the common starting points for the development of flood-response strategies is the delineation of flood hazard maps. Flood maps, showing the extent and depth of inundation for different magnitudes of floods, are useful in emergency situations as well as for flood zoning as previously described. These maps may be prepared beforehand for a range of events with different probabilities and the forecasting system should then relate to these maps, so that the expected extent of flooding can be visualised.

For flood hazard warning systems to be effective, the process must ensure that decisions are clear and transparent, specify the appropriate response, and the source of the decisions must be credible. An overall flood planning, warning and control decision-support system need to be defined to meet the needs of the end-users. The main stakeholder groups involved in the process should include operational managers responsible for urban stormwater system, their technical advisers, as well as other managers of other urban services (Todini 1999). Therefore, agencies responsible for emergency response strategies need to maintain links with local communities and inter-agency liaison, and communication procedures need to be working effectively.

It is essential to have a well co-ordinated and trained team to respond to emergency situations and to assist in the provision of temporary shelter, medical care, food distribution and water supply (see Figure 10.6). Organisation and training of

Box 10.3   Flood hazard map distribution in Japan

In Japan, each municipality is required by law to make a disaster prevention plan in order to prepare against earthquakes and floods. In 1994, the Ministry of Construction started to encourage municipal governments to publish flood hazard maps to be compiled by each municipality to be distributed to every home in flood-risk areas. However, since 2001, the production and distribution of the flood hazard map became compulsory by law. The aim of the flood hazard maps is to minimise flood damage by giving information to the residents about flood risks in their area, supplemented with other information useful for evacuation. The flood hazard map provides a visual representation of possible flooded water depths using contours on the map to show estimated maximum flood depths of flooding under different rain events.

To manage the production of the hazard maps, each municipality sets up a committee consisting of various institutional stakeholders and community representatives. The Ministry provides technical support to the municipality through this committee and some basic standards are recommended to each municipality such as for the colouring of floodwater depth, which should be the same throughout the municipalities in the same river basin.

In order to evaluate the effectiveness of the flood hazard maps, questionnaire surveys were undertaken amongst residents. Overall, the flood hazard maps were found to be useful as a source of information about floods amongst the community members especially in high-flood-risk areas and they also were found to be useful to promote a greater preparedness for flooding. However, in some cases, the effectiveness was found to be constrained by the fact that some residents were not aware of the existence of the flood hazard map or how to respond to it.

Source: Shidawara (1999)

Figure 10.6   Residents affected by flood carry drinking water home from an emergency distribution point operated by CARE Bangladesh during the 1998 flood in Dhaka. (Photo: Reproduced with kind permission of the Disaster Management Project/CARE Bangladesh.)

10.4.1 Flood warning systems

Before embarking on a project providing real-time hydrological information to the public, it is important to assess which information the public needs to have and which information local authorities require in the flood situation. Whereas the public requires information that is clear and concise to help them gauge the level of risk, local authorities require more detailed information to implement and co-ordinate a comprehensive flood-response strategy.

Therefore, before disseminating real-time hydrological information related to flood warnings it is important to consider:

(1) What type of hydrological information is valuable from the perspective of the public?
(2) How and where should the real-time information be presented to the public?

Detailed real-time flood forecast information is important for the local authorities to choose the best operations strategies for weirs, pumps and gates. This information will need to be at a much higher level than the details and accuracy of information required by the public. In this respect, it is therefore also important to consider what technology is required to provide real-time hydrological information as well as issues related to the capacity of service institutions to utilise this technology (Nielson 1998).

Figure 10.7 illustrates a typical structure for a real-time control system for monitoring and responding to flood risks. Once a flood event has been identified, the

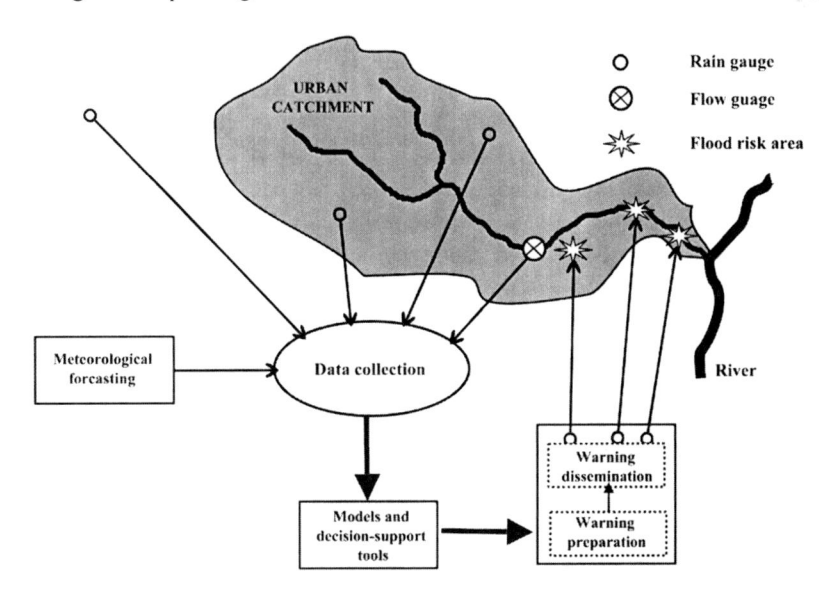

Figure 10.7 Typical forecasting and warning activities. (Adapted from Rowney *et al.* 1997.)

168   Urban stormwater management in developing countries

medium of communication is critical to ensure that mobilisation of communities and evacuation response strategies are effective. Public information sources such as radio may be used, but there are other approaches towards the dissemination of flood warnings that may be utilised such as the one described in Box 10.4. However, a combination of different communication strategies is advisable to ensure that as many people as possible get to know about the flood warning.

The example described in Box 10.5 describes the development of an integrated approach towards flood warning based on a real-time control system in combination with non-structural strategies.

Box 10.4   A real-time hydrological information system for Bangkok

In Bangkok, real-time rainfall data has been made available to the public through the Internet, hand-held personal digital assistants (PDAs) and mobile phones. The benefits include a public rainfall and flood information service, which may be used to inform residents about flood threats. In addition, the hydrological information can be applied in conjunction with real-time hydrological and urban drainage models providing decision-support and warning systems to deal with urban flooding and flash floods. The main benefit from this system is a public rainfall and flood information service, like the daily weather forecast and traffic information, about streets with a potential risk for flooding. A decision-support system for reducing flooding is currently being implemented by the Bangkok Metropolitan Administration for the part of the city located on the east side of the Chayo Praya river.

Source: Mark *et al.* (2002)

Box 10.5   Dynamic flood warning system: an integrated approach to disaster mitigation in Bangladesh

The Sundarganj sub-district (Thana) is located in the Gaibandha district in the northern part of the Bangladesh and is bounded by two major rivers, the Brahmaputra (Jamuna) on the east and the Teesta in the north. Due to the high rainfall intensities, the topographical conditions and the low gradient of the terrain, combined with insufficient capacity of hydraulic structures and obstructions in the drainage system, the area is subjected to floods almost every year. In order to develop an integrated approach to the management of flooding, the following components were developed:

(1) *Identification of areas of flood risk*
    Flood inundation maps were created using historical rain events and flood-risk areas were delineated using dynamic spatial modelling. GIS was used for the analysis to determine hazard zones in the maps, which serve as risk-zone identifiers. This information enables the communities that are

> most likely to be affected, and the most appropriate locations for the evacuation of the affected population to be identified.
> (2) *Real-time flood warning*
> The availability of accurate data before a flood event facilitates effective decision-making in adopting proper measures towards disaster preparedness, mitigation, control, planning and management. To develop an effective decision-support system for flood-risk assessment, it is important to apply the most efficient methods in flood forecasting and warning system associated with real-time data collection system. An integrated approach was developed using a hydrodynamic model (MIKE 11) and GIS to release warning of flood in advance of 72, 48 and 24 h.
> (3) *Flood evacuation routes*
> In order to develop an effective flood warning and evacuation system, it was necessary to identify the most efficient route to provide direction for the vulnerable people to go to a safe place as quickly as possible. The software ArcView Network Analyst was used to find the quickest way to reach the safest location, and by overlaying this information with population data, it was possible to assess which were the main evacuation routes and which were the main destinations of safety. This analysis enabled the assessment of the current provision for assistance during a flood-emergency situation.
>
> *Source*: Aziz et al. (2002)

10.5 FLOOD RECOVERY AND REHABILITATION

Post flood recovery involves action that is required after a flood event, which enables communities to respond to the consequences of floods, and allows urban authorities to organise and effectively co-ordinate relief activities. Rehabilitation involves the provision of services and facilities that will enable flood-affected families return to their homes and continue normal life as quickly as possible. As shown in Figure 10.8, this involves numerous decisions regarding return to normal life after a flood event related to:

- Evaluation of damages.
- Rehabilitation of damaged properties.
- Provision of flood assistance to flood victims.

Especially for the urban poor, the recovery from a severe flood event can consume considerable time, energy and resources to reconstruct damaged properties. It is therefore important that medium- to long-term assistance that supports the communities affected by flooding makes the best use of local resources and support from national and international relief agencies.

Figure 10.8 Flood rehabilitation measures (Andjelkovic 2001).

Flood insurance

Flood insurance is a financial mechanism that provides compensation for damage and loss caused by flood to residents and property owners who live in areas that are prone to flooding. It may provide an effective mechanism for providing support for families affected by flooding, enabling them to spread an uncertain but large loss over a long period of time, and also provide mechanisms of spreading flood loss over a large area and a large number of individuals (Andjelkovic 2001). However, it is often difficult to determine a realistic premium to build up a flood insurance portfolio based on accurate flood risk assessments and depth–damage relationships.

In addition, the insurance sector in developing countries is often not well developed and insurance agencies are not well deployed to deal with flooding. In general, only a handful of larger companies have access to sufficient capital to provide adequate insurance cover in the event of a large flood, which affects many people at the same time. These companies have the capacity to draw from the resources in other insurance sectors and from other areas, which have not been affected by the flooding.

In countries such as Bangladesh, it is unlikely that private companies will be able to provide adequate flood insurance in the near future, and consequently, property owners often have to purchase different policies in order to insure against all major disasters. The possibility for poor households who live in the highest-risk conditions to obtain an insurance policy is almost zero. In addition, where insurance is provided, there is the additional problem that claims may be artificially inflated due to the fact that residents may see it as an opportunity to make some money and submit exaggerated or false claims (Faisal *et al.* 2000).

10.6 REFERENCES

Andjelkovic, I. (2001) *Guidelines on Non-structural Measures in Urban Flood Management.* IHP-V Technical Documents in Hydrology. No. 50. Project IHP-V Project 7. UNESCO, International Hydrological Programme, Paris.

Aziz, F., Tripathi, N. K., Mark, O., Kusanagi, M. (2002) Dynamic flood warning system: an integrated approach to disaster mitigation in Bangladesh. *Proceedings of the International Conference,* 'Map Asia 2002', Bangkok, Thailand.

Campana, N. A. and Tucci, C. E. M. (2001) Predicting floods from urban development scenarios: case study of the Dilúvio Basin, Porto Alegre, Brazil. *Urban Water* 3(1–2), 113–124.

Faisal, I. M., Kabir, M. R. and Nishat, A. (2000) Non-structural flood mitigation measures for Dhaka city. *Urban Water* 1(2), 145–153.

Fordham, M. (2000) Managing floods in a changing social environment. *Proceedings of the Conference on 'Floods and Flooding in a Changing Environment'*, 28–29 April, University College Northampton, UK.

ISDR (2001) *Guidelines for Reducing Flood Losses.* International Strategy for Disaster Reduction UN International Strategy for Disaster Reduction – http://www.unisdr.org/eng/library/isdr-publication/flood-guidelines/isdr-publication-floods.htm

Mark, O., Boonya-Aroonnet, S., Quang Hung, N., Buranautama, V., Weesakul, U., Chaliraktrakul, C. and Chr Larsen, L. (2002) A real-time hydrological information system for Bangkok. *Proceedings of the International Conference on Urban Hydrology for the 21st Century*, 14–18 October, Kuala Lumpur, Malaysia.

Nielson, T. K. (1998) *The Technological Upgrading of Service Institutions.* Intermediate Technology Publications Ltd., London, UK.

Rowney, A. C., Stahre, P. and Roesner, L. A. (eds) (1997) *Sustaining Urban Water Resources in the 21st Century. Volumes 1 and 2; Conference Proceedings*, Malmo, Sweden.

Shidawara, M. (1999) Flood hazard map distribution. *Urban Water* 1(2), 125–129.

Stephens, C., Patnaik, R. and Lewin, S. (1996) *This is My Beautiful Home: Risk Perceptions Towards Flooding and Environment in Low-Income Urban Communities: Case Study in Indore, India.* London School of Hygiene and Tropical Medicine, London.

Todini, E. (1999) An operational decision support system for flood risk mapping, forecasting and management. *Urban Water* 1(2), 131–143.

Together Foundation and UNCHS – http://www.ucl.ac.uk/dpu-projects/drivers_urb_change/urb_infrastructure/pdf_city_planning/HABITAT_BestPractice_Disaster_Bogota.pdf. Last Accessed 13/11/04.

Tucci, C. E. M. (2002) Flood control and urban drainage management in Brazil. *Waterlines Journal.* Special edition on 'Urban Stormwater Drainage' April 2002, Intermediate Technology Development Group.

11
Participation and partnerships

The roles of different stakeholders and the importance of partnerships in urban stormwater management is a reoccurring theme throughout the book, which has been emphasised previously in discussions related to integrated water resource management (IWRM) (Chapter 3), policies and institutions (Chapter 4), and non-structural stormwater management practices (Chapter 10). Participation can take a wide variety of forms and the examples presented in this chapter illustrate ways in which local stakeholders may be involved in the planning, design and implementation of stormwater management strategies. Although the chapter aims to present a positive picture of participation, there are potentially many constraints associated with participatory processes, which may relate to a lack of motivation, conflicts of interest and a lack of effective communication and therefore understanding which can undermine the success of the project. It is therefore important to involve specialists who are experienced in participatory planning to ensure that participation is approached in the right way.

11.1 FORMS AND POTENTIAL BENEFITS OF PARTICIPATION

The traditional focus of urban stormwater management has been upon the provision of drainage infrastructure, and the responsibility for planning and design aspects

© 2005 IWA Publishing. *Urban Stormwater Management in Developing Countries* by Jonathan Parkinson and Ole Mark. ISBN: 1843390574. Published by IWA Publishing, London, UK.

has been the domain of engineers. In this situation, the involvement of professionals from other disciplines or from local stakeholders who are likely to be affected (for better or for worse) by the structural intervention is generally minimal or non-existent. However, as described in Chapter 4, the role of local stakeholders in the decision-making processes associated with the development of stormwater management strategies is increasingly recognised to be an important aspect of integrated urban water management.

Participation offers opportunities for urban authorities to assess the social feasibility of stormwater management systems and flood mitigation strategies. The involvement of local communities may help in the identification of problems and also in the planning of the layout of the drainage infrastructure itself. The knowledge and resources of local stakeholders may provide an invaluable contribution to the project design and benefits may be measured in terms of cost-effectiveness and increased sustainability of projects.

Community participation is generally seen to be an effective way of improving the delivery of basic infrastructure and services, especially in low-income, informal settlements. It is especially relevant when planning and designing projects for informal settlements as a way to involve the urban poor and marginalised groups in decision-making and urban planning processes related to infrastructure and service delivery for low-income communities (Imparato and Ruster 2003).

However, participation should not be viewed only as being relevant for low-income communities. Participation is also particularly important for the successful implementation of non-structural stormwater management and pollution control strategies, and should aim to involve a wide range of local stakeholders that have vested interests in the control of stormwater.

The extent of participation exercised in the planning process may vary considerably and, in the following list, the extent of participation increases towards the bottom of the list:

(1) Information giving (e.g. public presentations).
(2) Information gathering (e.g. opinion surveys).
(3) Discussion and negotiation in planning processes.
(4) Power-sharing and participatory decision-making (e.g. participatory budgeting).

There are a number of key principles that underlie successful participatory planning. These same principles, if not adhered to, may undermine the planning process and the potential benefits of working with local stakeholders will be lost.

11.1.1 Key principles of participation

(1) *Consensus*: The aim is to reach a consensus agreement taking into account the opinions of a wide range of local stakeholders. Participation can help to resolve conflicts through discussion, but it should be recognised that it will often not be possible to reach universal consensus. This means that it may be sometimes necessary to go against the wishes of some stakeholder groups for the sake of a common good.

(2) *Inclusive*: The participatory process should not exclude any member or group from the decision-making process.
(3) *Transparent*: Information to assist the participatory discussions and the resultant decisions and outcomes should be clear and understandable to all of those involved.
(4) *Accountable*: The way decisions are made during the planning process should be accountable, that is, in the future, those who made the decisions should be held responsible for their decisions.
(5) *Decisive*: At the end of the planning process, the final outcomes should be decisive so that there is no confusion about what the participatory process has achieved and the decisions that have been made.

11.1.2 Who should be involved?

The involvement of the general public in decision-making processes is a complex task for urban authorities, particularly as they have to cope with a wide diversity of stakeholders with differing perceptions of the problems and conflicting priorities for overcoming these problems. In particular, urban communities are generally heterogeneous and stakeholders show different perceptions and relationships to natural hazards according to their socio-economic and cultural backgrounds.

The emphasis on stakeholder involvement in urban stormwater-related policy-making has led to the need for the technical professionals to become proficient in facilitating such groups. However, it remains a challenge for urban stormwater engineers to work with diverse groups of stakeholders and this means that they need to work with people who have the necessary skills on social interaction and mediation.

Participation needs to consider the ways in which the roles of men and women may differ in decision-making processes and those responsible for urban stormwater management should pay close attention to the views of women. It is especially important to ensure that women's opinions are equally and proportionately represented, because they often lack ownership or control of resources, access to information and decision-making authority (Francis 2002). In some societies, where the woman's role in society is traditionally not respected as that of men, it is necessary to adopt planning exercises, which actively encourage the participation of women.

The presence of women in the planning process is particularly important for the development of workable flood protection and mitigation strategies. In addition, women are often more responsible than men for maintaining the quality of the environment in and around the house and will therefore have a direct interest to ensure that the drains that pass in front of their houses are free-flowing and clear from debris.

11.2 PARTICIPATION IN PLANNING AND DESIGN

The planning of a drainage project should be based on an understanding of the needs and expectations of the community that it is designed to serve. In principle, the project planning process should aim to identify the most appropriate solution that is prioritised according to local stakeholder demands, and their ability and willingness to pay for improved services. It is especially important to ensure that those affected are engaged in the planning process where it is considered necessary to relocate families who have constructed dwellings on drainage pathways or flood plains.

Physical data for planning and design may be scarce, especially in informal areas, and the available information may not cover the community where the project is to be carried out. In such cases, the community can help by describing where major flood problems occur and providing information about previous floods. In some cases, the community experience may not cover very big flood events, for example with a return period of 20–50 years.

Community members will also be important sources of information to confirm where the drainage problems are worst and to help develop a drainage plan that is accepted by the community and one in which community members will play their role in maintaining the system and keeping it clear from blockages. The best way to assess the extent to which flooding is perceived as a problem is to ask local residents. Although they may have little theoretical knowledge about hydrology and hydraulics, they will often intuitively know about the drainage problems in the area where they live. Based on past experience, particularly in well-established communities, local residents can often recognise the types of rainfall that are likely to lead to flooding. They may also have an inherent knowledge of the flooding characteristics and know about the problems that floods cause (see Chapter 2).

In order to gain a representative perspective of how local residents view flooding problems, it will be necessary to consult a wide range of stakeholders. As described above, there are a number of ways of approaching participation in the planning process. This may simply involve surveys to evaluate public opinion in relation to the impacts and benefits of the proposed plan. However, there are opportunities for a greater degree of involvement of the local population in directing the development of the plan, which can be encouraged through the course of discussion at neighbourhood public meetings.

Alternatively, a more detailed assessment may involve qualitative evaluation to assess perceptions of flooding in relation to other issues affecting livelihoods. Questionnaires may be developed to help focus and guide semi-structured interviews of local residents to learn more from them about their perceptions of flooding and the problems of poor urban drainage. This level of public participation can be time consuming and many projects cannot involve this level of detailed planning. Therefore, as it will not be possible to interview everybody, it is important that those who are selected are considered to be representative of the community.

A range of tools exist for participatory planning which can form the initial basis for the development of a partnership between local community and local government agencies. The institutional arrangement for a proposed environmental management plan (EMP) in Lucknow described in Box 11.1 stresses the importance of partnerships in aspects of urban environmental management related to the planning

> **Box 11.1 Environmental management of an urban catchment in Lucknow, India**
>
> A participatory process was adopted during the development of an integrated Environmental Management Plan (EMP) for critically stressed urban catchments in the city of Lucknow, Uttar Pradesh, India. The aim was to develop a solution to the problems of flooding and to provide services for the slum dwellers in low-lying areas adjacent to natural drainage paths. The catchments were identified based on discussions with local stakeholders to identify the most degraded urban rivers combined with secondary information from municipal agencies and pollution control boards on the quantities and quality of wastewater discharged. Once the catchments had been identified, various Participatory Rapid Appraisal (PRA) studies were undertaken in different locations within the catchment in order to gather primary information about the level of existing infrastructure and services. The main objective was to see how community members perceived and prioritised their problems, and this involved transect walks (see Figure 11.1) and various other participatory exercises involving local stakeholders to obtain an overview of the physical environment and its degradation in order to provide the following information:
>
> - Nature of settlements and the built up environment along the *nala*[1] banks.
> - Hierarchy of open spaces along the entire *nala* stretch and community's response to these open spaces.
> - Residents' behaviour towards waste disposal and its resulting impact on *nala* environment.
> - Profile on the type of activities conducted on the banks of *nala*.
>
> *Source*: Singhal and Kapur (2002)

and implementation of stormwater drainage. The project involved the application of participatory rapid appraisal (PRA) techniques in the formulation of a community-based EMP for a critically stressed urban catchment in India.

PRA and Beneficiary Assessment are two examples of qualitative research tools, which may be used to ascertain the views of intended beneficiaries. These methodologies use a variety of low-cost techniques such as interviews, focus group discussions and direct participant observation in order to promote an improved understanding about the opinions and concerns of local stakeholders. PRA studies should avoid leading questions or presumptions about the existing situation and it is also very important that these preliminary planning activities do not lead local residents to high expectations about the level of investment or scale of intervention, which may lead to future disappointments if these expectations are not reached.

---

[1]*Nala* is an Indian term for an urban drainage channel.

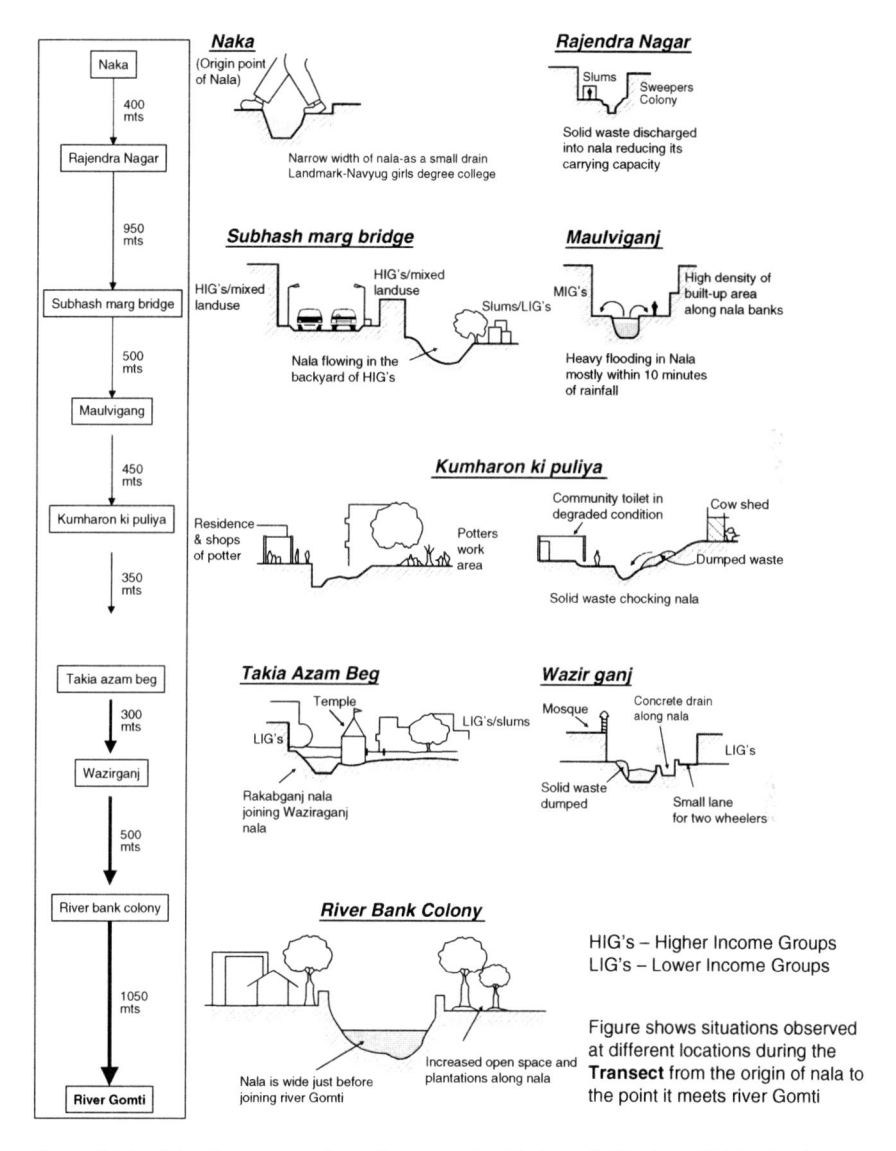

Figure 11.1 Visual representation of a transect walk through Wazirganj Nala, Lucknow, India. (Reproduced with kind permission of Shaleen Singhal and Amit Kapur.)

11.3 PARTNERSHIPS IN PROJECT IMPLEMENTATION

The conventional approach to project implementation for drainage construction usually involves the contracting of engineering companies under supervision by engineers from the local government agencies responsible for drainage in the city.

178   Urban stormwater management in developing countries

Figure 11.2 Construction of a drainage channel using local contractors. (Reproduced with permission of WEDC, Loughborough University.)

For larger-scale infrastructure, the need for specialist skills and equipment necessitates the use of contractors with the necessary expertise to undertake the work. Use of local contractors and labour for small-scale community-level infrastructure (see Figure 11.2) can reduce costs involving outside contractors and facilitates transfer of skills into the community. The involvement of local business and micro-contractors, especially those based within low-income communities, can be important as it generates economic activity and stimulates local interest in the project.

In many cases, community-based initiatives in both informal and formal settlements have already organised themselves to fill the gaps left by formal mechanisms of service delivery offered by government agencies. In these situations, there are a number of opportunities to involve local stakeholders in project implementation and urban drainage projects are amongst the range of urban services and infrastructure in which communities may be involved in the provision of local-level infrastructure.

As shown in Figure 11.3, for small-scale infrastructure projects, there are many opportunities for involvement of local stakeholders in drainage construction. These may include both local builders, masons and semi-skilled labourers, but it may also involve some community members as unskilled workforce (see Figure 11.4). Amongst the potential roles in the implementation of urban drainage projects, community members may be directly involved in the digging of drains and prefabrication of drain components in order to reduce the total cost of construction.

The community contracting approach involves the award of contracts for implementing infrastructure works to local community organisations or groups. The contract for works is directed towards the beneficiary groups and the competitive tendering process is avoided. In this approach, benefits of the contract go to the community and not to a contractor, middleman or development agency. The

Participation and partnerships 179

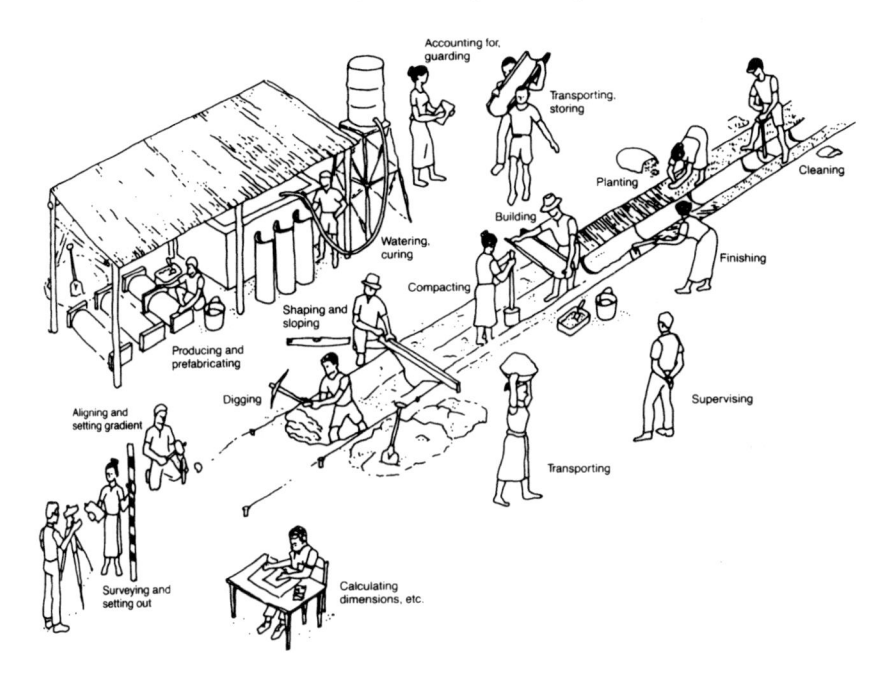

Figure 11.3 Tasks in drainage construction (Cairncross and Ouano 1991). (Reproduced with kind permission of WHO.)

Figure 11.4 Residents (women and men) involvement in community labour-based infrastructure improvement in Dar es Salaam, Tanzania. (Photo: Alphonce Kyessi.)

concept is promoted as a more efficient, appropriate alternative to expensive, top-down, contractor-driven urban upgrading projects (Cotton *et al.* 1998). An example of construction of stormwater drainage channels through community-based organisation approach is described in Box 11.2.

> **Box 11.2 Community contracting for drainage channel construction in Dar es Salaam, Tanzania**
>
> In Hanna Nassif, Dar es Salaam, through a project supported by the International Labour Organization, local community members organised themselves in groups and approached the municipal authorities responsible for infrastructure and services and offered their resources to improve the urban environment. The construction of the drainage system was based on community contracts, which employed the services of both men and women (see Figure 11.4). The community formed a Community Development Committee (CDC) with representatives from each of the six areas within the settlement (11 women and 8 men) and was divided into an Economic and Finance subcommittee and Construction subcommittee. The role of the CDC was to act as an intermediary between the community, the Dar es Salaam City Council and the donor community. The CDC was also responsible for mobilising the community to provide the financing for the infrastructure, to negotiate the terms of reference for the community contracts, and to agree upon the future management arrangements for operation and maintenance. The experiences highlight the fact that community participation is sensitive to mobilisation and awareness raising during the early stages of the project. In addition, the level of stakeholders' participation might be influenced by a number of factors and that the nature of external stakeholders may influence the level of the community participation. In Dar es Salaam, the nature of participation was primarily a result of the income-associated activities including engaged paid labour, and credit scheme incentives, which were provided as the project took off.
>
> *Source*: Kyessi (1997), Mulengeki (2000)

The potential benefits of community contracting and management include:

(1) *Increased value for money*: Those who have a vested interest in the outcome of an infrastructure improvement scheme are more likely than outside contractors to pay greater care and attention to the quality of the works.
(2) *Benefits retained within the community*: Community contracting not only provides improved facilities for poor communities. It also provides income for those community members who are involved in providing the facilities.
(3) *Improved liaison with local community members*: Conventional contractors may not be used to working in the closely confined conditions found in many low-income settlements. As community contractors are based in the community, they are much more likely to be sensitive to the concerns of local people.

However, more common is the involvement of community members as a collective group to help manage the construction. This group will be able to assist in the construction by overseeing the quality of the work and by mobilising the community to contribute towards the cost of construction. The group may also help to negotiate with the local contractors and the suppliers of building materials.

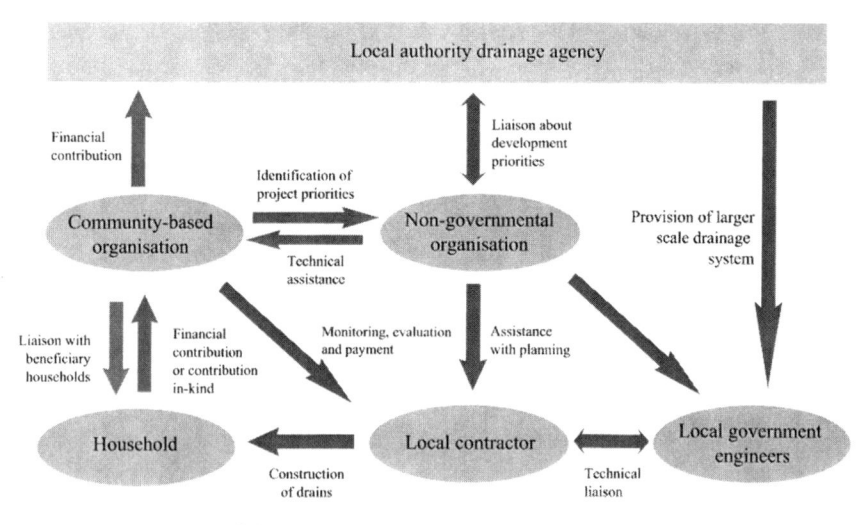

Figure 11.5 Process of drain construction using the community-based approach. (Adapted from Kyessi 2003.)

Figure 11.5 illustrates the various local actors involves in the community-based approach to provision of small-scale tertiary-level infrastructure, in which the community-based organisation plays the role of the drainage committee. Although there are benefits, this can result in excessive complications and high transaction costs can be incurred if the process is not managed well. A disadvantage is the need to train community organisations for higher levels of supervision and guidance than would be required with an experienced contractor. In addition, they require careful monitoring to ensure that process is not corrupted by local vested interests and personal gain.

Community organisation in these activities is critical to the success of the project and it will be necessary to establish a drainage committee to organise the community's contribution to the drainage project. This is most likely to succeed if it built around existing groups or institutions that are already established and recognised in the community as having some authority. This will depend much on the existing social structures and representative organisations. In some countries (such as in Vietnam) there is a very strong structure of community representation based on the People's Committees whereas in other countries (such as in Bangladesh) official political representation is much less prominent at the grass roots level, and communities reply much more upon local community-based groups and non-governmental organisations (NGOs).

It is important that one group in the community does not end up having all the responsibility for decisions. Although it is important for the group to have adequate representation, it should not be too large as this may result in becoming too difficult to reach consensus of opinion. It should be comprised of members from different parts of the neighbourhood and it is essential that the committee should include equal representation from men and women as well as principal ethnic and religious groups in the community.

Whatever the situation, the mechanisms of participation and social interactions are highly variable and complex, and often it is not possible to work directly with local-level stakeholders. It will therefore be necessary to promote a structure to encourage social interaction between the various actors involved in the project. In many projects involving community participation, local NGOs take on the role of the intermediary, in order to encourage liaison and communication between the different actors involved. NGOs often have considerable expertise in mobilising communities, organising fund-raising, as well as overseeing external technical assistance where appropriate to provide basic infrastructure, including drainage channels within their locality (Kyessi 2001; 2003).

11.4 PARTICIPATION IN OPERATION AND MAINTENANCE

There are a number of opportunities for drainage agencies to involve local organisations in activities related to operation and maintenance. This may involve the services of local private sector companies hired directly under contract. The role of the private sector for maintenance activities is potentially particularly important as it reduces the need for the municipality to employ a large staffing for these activities. Lease contracts offer clear incentives and opportunities to reduce costs by introducing competition, provided the cost of cleaning is not set by local government.

Alternatively, the responsibility for operation and maintenance may rest with a community-based organisation that may utilise internal resources within the community to respond voluntarily to operational problems. Regular inspection and cleaning of drains are important tasks that communities can undertake without specialised skills. Community-based organisations may contract local companies or individuals to undertake cleaning tasks. However, the organisation of community members who are ultimately responsible for actually carrying out the cleaning of the drains can be problematic and the key is therefore community organisation and the allocation of certain basic tasks to specific members (UNCHS 1986). In addition, unlike capital investments for construction, it is often more problematic to get communities to share maintenance costs over a sustained period of time.

It may be appropriate to contract the services of a member from the local community to be responsible for drain cleaning. This person may be contracted by a management committee who collects a small fee from the community to pay for the cleaning services. This may overcome problems of reliance on the active participation of all households in drain-cleaning activities who may perceive these activities to be degrading or unnecessary, especially where they do not suffer directly from any of the problems related to poor drainage.

However, there is a danger to assume that community members will automatically accept the responsibility for drainage system maintenance. In fact, many only resort to undertaking these kinds of task when then they realise that waiting for someone else to do it (generally the maintenance department) is going

Figure 11.6 Local community members in an informal settlement in George Compound in Lusaka to clear the stormwater drain next to their yard. (Photo: Martin Mulenga.)

to take too long. As a result, community members may take the initiative and clean the drains themselves (as shown in Figure 11.6), but to formalise these initiatives can prove to be difficult and it is wrong to expect that local residents will automatically take on this role.

The main problem with community-based maintenance is that tasks and their importance are not always immediately apparent to community members who may neglect preventive maintenance. In addition, the best efforts of the community may be thwarted if the drainage system affected by the operation of the downstream drainage system. This may be due to a lack of operation of irrigation sluices (managed by different institutions), backwater effects from surface waters, a lack of pumping and/or blockages in the system.

11.5 PARTICIPATION IN NON-STRUCTURAL FLOOD CONTROL STRATEGIES

Participation is particularly important in flood control strategies in order to address the diverse perspectives of flooding and to understand community responses.

184   Urban stormwater management in developing countries

Experiences in municipal-level disaster planning shows that authorities expecting public support for their local mitigation strategies should first let the public express its own definition and perception of risk (related to hazard, threat, danger, vulnerability, protection level, etc.) (Affeltranger 2001).

Professionals involved in flood planning and management employ a range of techniques, such as public meetings with slide and video displays, and written information or newsletters (Fordham 1999). However, more intensive participatory studies can offer greater insight into how to design effective non-structural flood control strategies.

Participatory planning is particularly important to understand the differences in vulnerability, responses to flood warnings and understanding of the flood warnings, which may result in different levels of impact upon different social groups and subsectors of society (Affeltranger 2001). Participatory processes may be of use at different stages of the disaster mitigation cycle (see Figure 11.7).

There are several reasons in support of public participation in disaster planning and emergency management:

(1) Preparing the community before the disaster may reduce damage.
(2) A larger number of lives may be saved by a trained community before the arrival of external relief forces.
(3) Survival and health problems are better managed if the community is active and organised.

In addition, a well-prepared community may also contribute to enhance the quality of external relief aid by providing information on disaster damage, by contributing to proper assessment of needs, and by ensuring appropriate distribution of aid. This knowledge may relate to the temporal and spatial distribution of past flooding, but there are limitations to the scientific validity of community-borne hazard analysis data. For example, flooding may be often be exaggerated by community members who hope this will be a way to secure additional investment

1 - **Disaster management** (emergency and relief)

2 - **Rehabiliation and reconstruction** (structural and non-structural)

3 - **Feedback** on disaster management; analysis of community response

4 - **Mapping** of hazards and vulnerability analysis

5 - **Strategic planning** by the municipality for disaster management

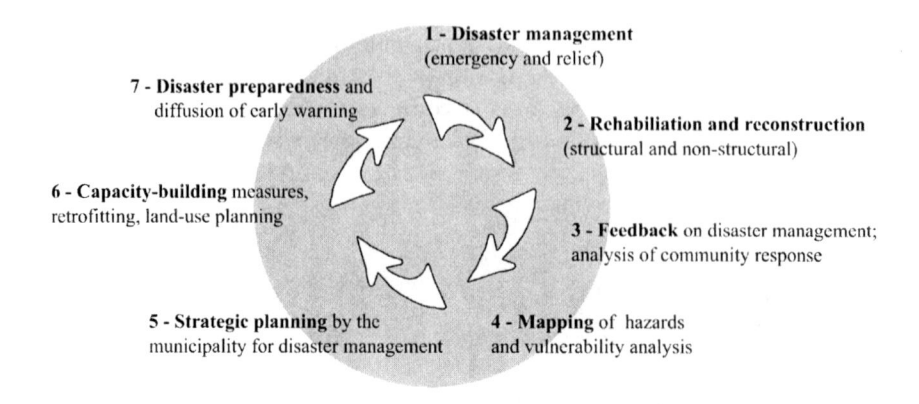

6 - **Capacity-building** measures, retrofitting, land-use planning

7 - **Disaster preparedness** and diffusion of early warning

Figure 11.7   The disaster mitigation cycle. (Adapted from Affeltranger 2001.)

in infrastructure in their area, or underestimated as a result of concerns that land value may decrease after hazard maps have been made public. For this reason, involvement of the public requires efficient scientific guidance, as well as expert validation of empirical data produced (Affeltranger 2001).

Participation of women is important for identifying the impact of floods based on their specific roles and responsibilities. The role of women during the planning and design of flood mitigation strategies is essential and women are often key players after disasters. The disruption of normal life and the pressing need for rapid recovery present new possibilities for women to overcome traditional barriers. Focusing solely on women's vulnerability may be misleading since they often

---

Box 11.3 Role of women in the design of flood risk reduction and disaster mitigation in Jaleshwor municipality, Nepal

Jaleshwor municipality, in partnership with Nepal Red Cross Society, has been implementing low-cost risk-reduction measures since 2000 where communities have faced tremendous problems due to the fact that they live in high risk conditions caused by flooding. The project was implemented in four pilot wards through community participation, and from an early stage focused on the importance of women in the planning process as well as in the implementation of urban disaster management programmes. During the planning stages, the project adopted PRA methods and tools to identify the vulnerability problems that women face during flood events, and to understand more about their perceptions of these problems with a view to developing appropriate interventions. The PRA tools were used extensively in assessment exercises involving the active participation of women with the aim to sensitise them towards needs identification, situational analysis and problem-solving processes.

This exercise led to an increased awareness of the existing problems in the community and the involvement of women was particularly beneficial in order to gain insight into the local knowledge about the conditions of life in flood risk areas. However, in spite of the important lessons from these projects, women are still generally viewed as beneficiaries and the majority of urban practitioners in Nepal do no recognise the importance of their role in the planning and decision-making process for the design of development of risk reduction and disaster management strategies. Therefore, greater emphasis needs to be placed upon awareness raising amongst local politicians, government officials as well as civil society organisations to gain a commitment towards collective action, which is needed to implement policies and practices which provide equal opportunity to both men and women. As a result of these experiences, the greatest challenge and focus of future activities is upon the integration of the women in participatory planning practices in mainstream policies and programmes for disaster management.

Source: *CARE Nepal Newsletter*, Vol. 8, No. 1, CARE Nepal, Kathmandu

have coping strategies and knowledge that can be used to minimise the impacts of crisis, environmental change and disaster. This is often the result of women translating skills acquired through their daily routines into invaluable disaster assistance (Francis 2002) and is important for developing mitigation, coping and recovery strategies that will be targeted based on gender perspectives.

Specific actions are required for undertaking a gender perspective to flood management and these entail:

(1) Examination of gender roles and responsibilities in flood prone areas.
(2) Assessment of the extent and type of damage that floods are likely to cause and identification of the differential impact of floods on men and women based on their specific roles and responsibilities.
(3) Development of protection, coping and mitigation strategies that will be targeted based on the gender perspective.

The experiences described in Box 11.3 from Nepal provide a good example of how women may be involved in the development of flood-response strategies.

11.6 REFERENCES

Affeltranger, B. (2001) *Public Participation in the Design of Local Strategies for Flood Mitigation and Control.* Technical Documents in Hydrology No. 48. International Hydrological Programme, UNESCO, Paris.

Bhattarai, S. and Neupane, B. (2001) Informed decision making for drainage management. *Water, Sanitation and Hygiene: Challenges of the Millennium, Proceeding of 26th WEDC Conference*, Dhaka, Bangladesh, Water Engineering and Development Centre (WEDC), Loughborough University, UK.

Cairncross, S. and Ouano, E. A. R. (1991) *Surface Water Drainage for Low-Income Communities.* WHO/UNEP, World Health Organization, Geneva, Switzerland.

Chavez, R. (2002) Barrios El Café, La Mina and Hermanas Mirabel. In: *Santo Domingo: A Best Practice in Urban Environmental Rehabilitation.* Urban Notes Thematic Group on Services to the Urban Poor. No. 4 November 2002, The World Bank, Washington, DC.

Cotton, A. P., Sohail, M. and Tayler, W. K. (1998) *Community Initiatives in Urban Infrastructure.* Water, Engineering and Development Centre, Loughborough University, UK.

Fordham, M. (1999) Participatory planning for flood mitigation: models and approaches. *Australian Journal of Emergency Management* **13**(4), 27–34.

Fordham, M. (2000) Managing floods in a changing social environment. Paper presented at the *Floods and Flooding in a Changing Environment Conference.* University College Northampton, 28–29 April 2000.

Francis, J. (2002) *Implications of Gender in Floods.* Gender and Water Alliance. November 2002. http://www.genderandwateralliance.org/reports/discussion_paper_on_gender_ and_ floods_by_JF.doc Last accessed 4th November 2004.

Imparato, I. and Ruster, J. (2003) *Slum Upgrading and Participation – Lessons in Latin America.* Directions in Development. World Bank. Washington, DC.

Kyessi, A. G. (1997). Environmental hazards management – constructing stormwater drainage channels through community-based organization approach – case of Hanna Nassif unplanned settlement. Dar es Salaam, Tanzania. *Proceedings of a Conference on The Challenge of Environmental Management in Metropolitan Areas*, 19–20 June, University of London.

Kyessi, A. G. (2001) Community-based urban water management under scarcity in Dar es Salaam, Tanzania. *Proceedings of a Symposium 'Frontiers in Urban Water Management: Deadlock or Hope?'*, 18–20 June, Marseille, France. Edited by José Alberto Tejada-Guibert and Čedo Maksimović. IHP-V Technical Documents in Hydrology No. 45. UNESCO, International Hydrological Programme, Paris: 46–54.

Kyessi, A. G. (2003) Infrastructure improvement in informal housing areas; case of Buguruni Mnyamani, Dar es Salaam, Tanzania. *Paper Presented in an International Conference*, Montreal, Canada.

Mulengeki, E. F. (2000) *Potentials and Limitations of Community-based Initiatives in Infrastructure Provision in Tanzania*. Baruti-Kilungule Development Association (Bakida) and Hanna Nassif Community Development Association (CDA), Dar es Salaam, Tanzania. Department of Land and Water Resources Engineering, Sweden.

Singhal, S. and Kapur, A. (2002) Environmental management plans for the communities of Lucknow. *Waterlines* Special edition on 'Urban Stormwater Drainage' April 2002, ITDG Publishing.

UNCHS (1986) *Community Participation and Low-Cost Drainage – a Training Module*. United Nations Centre for Human Settlements, Nairobi, Kenya.

12
Economics and financing

Urban drainage infrastructure is expensive, especially in terms of capital investment but also for maintenance costs. It is especially expensive as it is only used to its full capacity infrequently and for relatively short periods of time. Therefore, relatively minor changes to engineering design parameters such as frequency of flooding can have a major implication on the size of infrastructure and therefore cost. Many local governments in developing countries face chronic financial problems, affecting their ability to extend services to underserved areas, and to operate and maintain existing infrastructure. In addition to lack of resources, many of them face problems associated with inefficient administrational and accounting procedures, and are often reluctant to fund operational and maintenance costs of urban drainage system due to the competition for funds. The seasonal nature of flooding means that the investment requirements for drainage are often sidelined due to the demands from other sectors. This chapter describes ways to address these issues and introduces various approaches towards financing and cost recovery that may be applied to urban stormwater management.

12.1 URBAN DRAINAGE – A PUBLIC GOOD

In the majority of cities, local government authorities are responsible for the provision of a wide range of urban services to urban citizens to enable them to live healthy and gainful lives. As well as flood protection and drainage of wastewaters

© 2005 IWA Publishing. *Urban Stormwater Management in Developing Countries* by Jonathan Parkinson and Ole Mark. ISBN: 1843390574. Published by IWA Publishing, London, UK.

> **Box 12.1 Characteristics of public and private goods**
>
> The question of what is a public good is best answered by considering the characteristics of private goods. From an economist's perspective, these are goods that can be offered in the market and traded between individuals or businesses. In the situation of an open market, which does not have any barriers to competition, there is scope for many producers and the purchaser has a choice about which product to buy. There is opportunity for negotiation in relation to the terms of transaction and, once the goods have been traded at the agreed price, the ownership of good is subsequently transferred to the purchaser. Thus, private goods have clear ownership and few others can benefit from their usage without the owners consent. Public goods, on the other hand, exhibit opposite qualities. They have shared ownership and, in principle, everybody – both the rich and the poor – has the potential to benefit. However, this does not necessarily mean that all people will benefit equally all of the time, it is important to recognise that public goods are not necessarily valued by all in the same way and that the priorities of different population groups may vary considerably.

and services for environment health, other services include roads and public transport systems and a range of other social services such as health and education.

The provision of urban drainage involves the provision of a public good. As described in Box 12.1, the differentiation between public and private goods is important and operates at a number of levels and this has implications on the provision of urban drainage in relation to the following:

(1) Ownership of the urban drainage infrastructure itself. In the majority of cases, even when the operator is a private entity, the ownership of urban drainage systems remains in the public domain.
(2) Management responsibilities for operation and maintenance services as well as issuance of flood warnings when the capacity of the system is exceeded during times of excess rainfall.
(3) Financing of urban drainage is complicated by the fact that residents who live in upstream areas are not affected by the flooding and generally perceive little direct benefit from the investments in downstream infrastructure.

However, although the benefits of public goods are not necessarily shared equally, urban drainage provides benefits for the city as a whole. These factors influence both the financing and cost recovery mechanisms, and the type of organisation that is responsible for provision of the good or service. Whereas market institutions are observed to meet demand via the supply of private goods and services, supply mechanisms for public goods are predominantly via political institutions and the responsibility for provision and management of urban drainage generally remains the responsibility of the public sector. Thus, the traditional financing mechanism for urban infrastructure involves public sector financing in which the drainage system is funded as a part of the municipal budget and costs are generally subsumed into the municipal taxation.

12.2 MUNICIPAL BUDGETING AND ACCOUNTABILITY

An overall lack of finances will invariably affect the amount of funds available for local authorities to allocate towards expenditure on improvements in urban drainage facilities. Urban drainage is one of a wide range of public services that competes for investments from the municipal budget. The decisions about how to allocate available resources are generally made by local politicians and civil servants working for local government administration. However, their decisions are not always responsive to public needs and do not necessarily finance the sectors where investment is most needed. This is particularly due to political patronage and sometimes misappropriation of public funds.

The efficacy of these organisations will relate closely to their ability to manage finances effectively. According to UNCHS (1993), common problems related to financial management that influence the financial viability of municipalities and their ability to provide effective services for urban communities include:

(1) obsolete accounting systems and practices, which provide minimum information for financial management;
(2) faulty budgeting systems and practices;
(3) gaps in transfer of central government funds to authorities responsible for maintenance;
(4) limitations of central government regulation;
(5) lack of performance orientation in local service delivery.

Although not always present in all societies, civil society organisations and pressure groups may form an important collective action by the citizens to lobby local government officials and demand better services. However, poorer communities on their own generally have limited influence and often do not benefit from local government spending. But, these same communities are often those that are most affected by the impacts of poor drainage and flooding.

One of the most common ways in which communities assess the effectiveness of local government is in its ability to provide public services. Willingness to pay studies (described below) may offer one way for local authorities to assess public perception of municipal service and opinions about future investment strategies. An alternative approach, which is not based directly on trying to put a cost to a hypothetical market, is to assess service providers using report cards as described in Box 12.2. The use of scorecards can have a strong influence on municipal budgetary systems and a way to enhance accountability.

The participatory process has been taken a step further in some cities in Brazil, where the decision-making process regarding prioritisation of a proportion of the

Box 12.2 Participatory evaluation of urban services using scorecards in Bangalore, India

The concept of governance scorecards was pioneered in 1993 by the Public Affairs Centre in Bangalore, India and has proved to be an effective mechanism for soliciting feedback from citizens about the performance of public service providers. Since then, scorecards have been prepared in collaboration with committed local community-based organisations from the cities of Ahmedabad, Calcutta, Chennai, Delhi, Mumbai and Pune. The majority of these scorecards have focused on urban public services in order to demonstrate how the feedback from civil society surveys can assist local governments to improve the delivery of public services. In addition, scorecards have also been prepared for specific sectors like the health sector as well as other issues such as electoral transparency and integrity.

The governance scorecard attempts to assess the quality, efficiency and adequacy of governance from the recipients' point of view and the main objectives are to:

- Obtain feedback from local residents about the quality and adequacy of public services.
- Identify what services people expect from the public sector and which from the private sector and how much they would be willing to pay for each type of service.
- Identify areas in which citizens experience problems in accessing the service and estimate the hidden costs incurred by the citizens.
- Evaluate the efficiency and effectiveness of service providers to respond to complaints.

One of the key advantages of the scorecard system is that it encourages citizens to adopt a pro-active stance by demanding accountability, accessibility and responsiveness from service providers. It may promote the participation of low-income and marginalised sectors of society to play a role in decision-making process and increase public awareness and generate collective action and pressure from civil society against poor service delivery and political patronage.

Source: Paul 1998

municipal expenditure and investment in improved services has been handed over to local communities (see Box 12.3).

12.3 DEMAND AND WILLINGNESS TO PAY

The perceived benefits of urban drainage, in combination with a wide range of other factors, can have a significant influence of communities' *willingness to pay*, and subsequently the ability of the service provider to recover costs from the provision of infrastructure and associated services to reduce flooding.

> **Box 12.3 Participatory budgeting in Brazil**
>
> Participatory budgeting is one of the most significant innovations in Latin America for increasing citizen participation and local government accountability. Citizen assemblies in each district of a city determine priorities for the use of a part of the city's revenues. Participatory budgeting allows residents to decide on investment priorities in their communities and to review government responses. This involves a sequence of public hearings for each neighbourhood and for the main city projects which may subsequently influence the decision to invest in urban drainage in relation to other perceived needs. In Porto Alegre this budget decision process has been developed for 12 years and in Belo Horizonte for 6 years. In the most recent one in Belo Horizonte, the Urban Drainage Plan has been chosen as a fifth priority among 34 main city projects.
>
> *Source*: Souza 2001; Tucci 2001

Cost recovery in developing countries is particularly complicated due to the fact that informal settlements can constitute such a large proportion of the total urban population. As these settlements are not formally recognised by local government as part of the city and may be illegally occupying government land, local councillors and politicians often argue that they should not be there in the first place and they are not entitled to urban services. In some countries, legislation precludes the investment of public resources in illegal or unofficial settlements. At the same time, the same households do not receive bills for services and the fact that they do not pay is often used as an argument that the poor communities are not able to pay for services. Consequently, a viscous circle of revenue deficit combined with increasing demand for infrastructure and services places an increased burden on urban authorities.

However, it is often overlooked that these communities contribute to the economy of the city as a whole and many urban economies are dependent on the low-paid workforce who live in these informal settlements. In addition to this, and of specific relevance to the provision of urban drainage, is the fact that it is not necessarily true to assume that poorer communities are unable or unwilling to pay for essential urban services.

There is increasing evidence to suggest that the urban poor are willing to pay for urban services, but this will depend upon the type and quality of service provided. Various other factors such as trust in local government and the perception as to whether it is a transparent organisation, which utilises resources efficiently, effectively and equitably will also be significant.

As described in Box 12.4, when poor urban communities in Zimbabwe were consulted about their willingness to pay a service charge for urban drainage, the resident expressed a very low willingness to pay as they perceived it to be the responsibility of local government. In this survey, the contingent valuation methodology (CVM) was used to estimate perceived benefits and costs of different proposals, taking into account both tangible and intangible impacts. The CVM

> **Box 12.4 Demand and willingness to pay for drainage and other sanitation services in Zimbabwe**
>
> Gutu and Gokwe are two low-income settlements in Zimbabwe. In order to help local authorities to make informed decisions about how much to charge for services, a study based on the contingent valuation method (CVM) was undertaken to calculate the costs and willingness of households to pay for urban services. Participatory approaches were used to measure willingness and ability to pay for services for refuse collection, sanitation and drainage. In order to assess willingness to pay (WTP), households were presented with a number of scenarios for improved wastewater and stormwater drainage services. In addition to the construction of facilities, they were then asked how much they would be willing to pay for the cleaning and maintenance of the drainage facilities. One quarter of interviewees refused to offer WTP bids – some of these said that they would want to see the facilities constructed first, before they would be willing to suggest how much they would pay. An additional third said that they were not willing to pay anything at all for urban drainage. The majority of interviewees cited a lack of trust in the local authority as the major reason, or they thought that drainage was the sole responsibility of the local or central government and expected these facilities to be constructed and maintained by the government at no extra cost to the household.
>
> The responses were very different for sanitation and solid waste management. WTP for these were comparatively high, although the survey showed great variation in the respondents' bids and the amount offered was still relatively low – both in absolute terms and as a percentage of household income. Although, WTP for urban drainage and wastewater disposal and treatment was very low, the results indicated that local authorities can improve their revenue significantly by charging an amount which households are WTP. In addition, it was shown that income has an effect on WTP, which suggests that environmental sanitation programmes should be linked to poverty alleviation strategies and be tackled as an integral part of housing and urban upgrading programmes. The responses also indicated that accountability and transparency are important aspects of service provision, which can significantly affect WTP. Local authorities can improve their financial position by charging tariffs, which are based on the actual cost of providing environmental sanitation services and the willingness of households to pay for these services.
>
> *Source*: Manase *et al.* (2002)

is based on a questionnaire to ask how much residents are willing to pay for improved services. However, it remains a controversial method as there are some underlying assumptions, which mean that it is not always appropriate. In particular, residents need to be familiar with the problems and proposed solutions to be able to assign a meaningful value. Its application is also limited as it creates a hypothetical market in which people answer questions related to the amount they are willing to pay for a theoretical good or service. Thus, the results from CVM surveys do not always necessarily directly relate to the amount that households will pay once the service is provided. However, it may be used to quantify the level of public support to assess the collective demand for different goods and consequently which ones are perceived to be more important.

The example described above illustrates a number of reasons why willingness to pay for drainage is often low. However, it is important to note that willingness to pay, as measured by contingent valuation studies, does not necessarily directly equate to demand for improved services. Demand for services is an important factor that needs to be considered in the provision of public services such as drainage infrastructure. The demands for improvements in urban drainage will be highly seasonal in relation to the onset of the wet season when the awareness of the need to invest in drainage infrastructure often increases and this provides support for investments in urban drainage and stormwater management programmes.

12.4 COSTS OF STORMWATER MANAGEMENT

Unlike other components of urban infrastructure such as water supply and wastewater management, urban drainage systems consist primarily of static engineered infrastructure and little in the way of machinery or other hardware that requires daily monitoring and supervision. The financial costs associated with urban drainage are summarised below:

(1) *Infrastructure construction*: design and implementation of new infrastructure, contracting of services of consultants and contractors.
(2) *Operation and maintenance*: cleaning of drains and disposal of waste, purchase of equipment, staffing costs.
(3) *Administration and management*: staffing and other costs associated with administration and management of the organisation.
(4) *Information management*: data collection and development of information management systems and computer simulation tools.
(5) *Public relations*: public information and awareness-raising campaigns.

Table 12.1 lists the costs of drainage and summarises the benefits that are likely to be realised due to these investments. Some of these have direct and tangible impacts whereas other effects are less tangible in economic terms.

The costs of stormwater systems vary according to many factors, but primarily these related to design capacity, which is based on storm return frequency. The volume of runoff is directly linked to the rainfall intensity and the proportion of the impervious area. A simplified relationship based on Heaney *et al.* (2002) is presented in Figure 12.1, which shows that the cost of service provision is related to the density of urban development as well as the flood return period specified by the design. It is noticeable that the cost function does not exhibit a linear relation-

Table 12.1 Costs and benefits of drainage and flood control systems.

	Costs of drainage	Benefits
Tangible	• Construction • Non-structural programme • Relocation • Administration • Insurance subsidy • Land acquisition	• Reduced flood damage • Land value enhancement • Reduced traffic delays • Reduced business losses • Reduced clean-up and relief costs
Intangible	• Environmental impacts • Health impacts • Social costs • Psychological effects	• Reduced inconvenience • Increased security • Reduced health hazards • Improved aesthetics

After Grigg (1987).

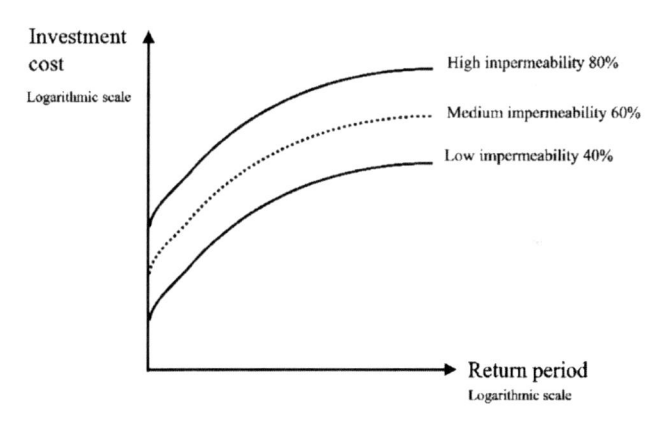

Figure 12.1 Costs of drainage infrastructure in relation to flood return period and impermeability of the catchment (after Heaney et al. 2002).

ship to return frequency and this may be related to the fact that many of the costs are not directly associated with the size of the drainage channels themselves and there are many fixed costs such as labour, land acquisition which will not change much as the return frequency is increased.

Prioritisation of investments according to the most important locations in the urban drainage network can dramatically reduce the need for large-scale upgrading of the whole system. In the UK, the Sewerage Rehabilitation Manual (WRc 2001) defines the urban pollution management (UPM) methodology (see Chapter 5) to identify critical sewers, which are sewers that have the greatest impact on failure in terms of the disruption to transport systems and structural damage. As described in Chapter 8, computer simulation models are important tools, that enable critical sewers to be identified and various improvement options to be analysed. In some cases, the hydraulic model may be linked directly to the cost estimation model, which provides a rapid comparison of various designs.

Financial costs may be reduced significantly by reducing the return frequency of flooding used in the design, which reduces the size of drainage channels. But this will reduce the flood protection and increase the flood cost. Other ways to reduce the scale of investment include.

- *Use alternative construction materials*: The most common type of material for the construction is concrete. However, drainage channels may be constructed from a range of materials and there are cheaper options such as open grassed swales (see Chapter 7).
- *Source control*: The source control strategies described in Chapter 7 can significantly reduce the need for large-scale downstream drainage infrastructure. As these benefits will be most noticeable to the local authority in terms of reduced investment requirements, the authority may consider offering grants or tax subsidies for those who use source control technologies and runoff reuse in their house.
- *Improved solid waste management*: Improved solid waste management can be more efficient and cost-effective way to maintain a drainage system than relying on drain cleaning. The focus on solid waste management should place greater responsibility on households to dispose of wastes more responsibly.
- *Private sector participation*: Private sector participation should not be confused with privatisation. The overall responsibility for system management may remain in the public sector, but there is considerable opportunity to enlist the services of the private sector for a wide variety of tasks. This can reduce costs by competitive bidding for contracts for construction or operation and maintenance.
- *Community participation*: As mentioned in Chapter 10, there is scope for reducing construction costs via community involvement in the form of unpaid or cheap labour. The potential advantage of community participation includes a reduction in financial burden on the central authorities because if residents are involved it means that they have an incentive to ensure that funds are used more efficiently. But, there are issues related to equity in community involvement such as those in the situation whereby higher income areas receive drainage out of the general taxation budget, whilst poorer neighbourhoods are expected to contribute unpaid or cheap labour.

12.4.1 Estimation of costs associated with flood damage

The estimation of flood damage is a fundamental step in the economic analysis of flood mitigation strategies. Empirical relationships between flood damage and frequency of flooding may assist decisions about expenditure and investment. Most damage models estimate monetary damage based on the type or use of the building and the inundation depth. Usually, building-specific damage functions are developed by collecting damage data in the aftermath of a flood. However, it

> **Box 12.5 Estimation of flood damages in Bangkok**
>
> A simple empirical model was developed to provide information for the city planners concerning cost estimation of the flooding and the return of the investment from implementation of alleviation schemes. For the Bangkok flood modelling study, flood inundation maps were overlaid on property maps of the city. A cost function for flood damage per establishment (in Thai baht) for the Sukhumvit area in Bangkok was developed to be:
>
> Flood damage cost $= a + b \times d + c \times T$ (baht)
>
> where d the depth of flood (cm); T the duration of flood (days); and a, b, c are the following function parameters estimated based on previous flood damage for different land use types in Bangkok:
>
> | Land use type | a | b | c |
> |---|---|---|---|
> | Residential | −300.5 | 45.4 | 33.5 |
> | Commercial | −2.2 | 88.1 | – |
> | Industrial | −1739.9 | 522.8 | 180.5 |
> | Agricultural | −1047.2 | 553.5 | – |
>
> *Source*: Tang *et al.* (1989)

is also feasible to undertake hypothetical scenarios to calculate the expected damage and costs for repair and recovery under different flood conditions (Smith 1994). Simplified relationships between damage and flood characteristics may be derived using observed and synthetic data and frequency damage relationships established between the hydrological and hydraulic parameters and damage caused by inundation. An example of an application of this approach is described in Box 12.5.

These types of depth-damage functions are widely adopted as the basis for assessing urban flood damage. However, although this type of approach is necessary for comparing mitigation measures and proposed interventions, there are inherent limitations of the accuracy of the methodology due to various simplifying assumptions related to the hydrological–hydraulic model approximation and the precision of the economic estimations. These assumptions may subsequently lead to errors in the damage estimates and thus the estimated benefits from proposed protection measures (Oliveri and Santoro 2002). Whilst it is possible to obtain reasonably accurate estimates of direct flood damages costs from survey and analysis of flood damage claims, the estimation of both indirect and intangible costs (see Chapter 2) is far more problematic. These impacts may be long lasting and it is almost impossible to measure them in financial terms (UNCHS 1993). As a result, it is difficult for agencies responsible for stormwater management to develop appropriate charging and systems for cost recovery and this remains a problem in many cities.

12.5 REVENUE GENERATION AND COST RECOVERY

The capacity of local governments to raise sufficient revenue from general tax revenues is often limited and many local authorities rely upon subsidies from central government and other sources of external financing – either from central government or international funding agencies.

Due to the funding requirements for large-scale infrastructure, there is often a need for external financing – in the form of a grant or a loan. There are opportunities for the private sector financing for construction which may require funding through long-term borrowing such as municipal bonds. Although this is often necessary, the reliance on external funding can perpetuate financial problems at the municipal level and distort priorities towards new investment and away from maintenance of existing infrastructure. It may also avoid the focus of attention on some of critical issues related to improving the operational performance via improved management practices, combined with a more rational use of resources.

Any charging mechanism needs to be fair (relative to other sources of revenue such as property tax) and needs to be based on equitable principles according to the ability of people and organisations to pay. According to Lindsey (1990), the degree of public support for charges will depend upon whether people support the scope and cost of programmes required to achieve objectives. This will be dependent on the perceived equity of the service.

A number of alternative stormwater funding sources described below may be considered which may offer opportunities, either individually or in combination, to provide funding for local governments to finance stormwater management programmes.

12.5.1 General taxes – local taxation

The general tax fund is the most commonly used source of funding for financing investments in stormwater infrastructure as well as ongoing operation, maintenance costs, and management overheads, both in developed and developing countries. Costs associated with construction, operation and maintenance of the storm drainage infrastructure are recovered through general tax revenues. Due to the fact that drainage is a public good, there is a strong basis for the argument that the most appropriate form of funding is taxation as oppose to fee or service charge. But, there is a danger that drainage systems can become neglected as the required investments are redirected towards other priority investment needs.

12.5.2 Permit fees

Permit fees are financial instruments for charging developers for the costs associated with the provision of drainage for new buildings. These funds are important sources of funding for new infrastructure and the associated costs may be passed on, either directly or indirectly to the user. However, these are not appropriate for raising revenue for operation, maintenance or retrofit of existing infrastructure and can only be seen as a complimentary source of funding. This is particularly true due to the fact that the revenue sources will be reduced during times when there are less new developments (Livingston *et al.* 1997).

12.5.3 Service charges

An alternative approach to the general tax is the *service charge* or *utility fee* in which the utility providing the service requests payments directly from individual property owners or, in the case of larger private developments, funds for projects in a district can be raised by assessing fees to landowners in the district. The service charge may be fixed according to the rateable value of the property in the same way that the taxation system operated, but for urban drainage, it is more pragmatic to adopt a tariff system which is determined according to the contribution to runoff from each property.

The amount of runoff may either be based on a land use classification system (as in the example from California described in Box 12.6) or it can be based on the area of the property and the proportion of impervious cover. This approach may provide direct economic incentives to property owners to reduce the amount of

---

Box 12.6   User fees – the key to managing stormwater costs

The city of Santa Cruz, California has found a cost-effective way to finance the costs of complying with recently mandated stormwater management regulations. To pay for increased stormwater-related costs, the city is adopting users' fees as an alternative to traditional municipal financing mechanisms. To generate the revenue needed to maintain the stormwater management programme, the city will charge residents a user fee based on the city's expenditures for stormwater-related activities. Using this approach, cities consolidate their stormwater and flood control activities into a stormwater enterprise and recover the costs from the residents who use and benefit from the system. Based on the criteria that each property inside city boundaries benefits from a stormwater pollution control programme, stormwater costs are allocated according to the benefit received, or the portion of stormwater that runs off from each plot (see Figure 12.2). This method of allocating costs is commonly employed by special flood control districts and stormwater utilities, and enterprises nationwide. Using a combination of land use runoff factors and coefficients developed by the city, the percentage of runoff from representative plots for each land use within the city was calculated. The city's imperviousness factors were developed from aerial photographs and data supplied by the tax assessor which were used to calculate the ratio of impervious area to total plot size.

Source: Null 1995

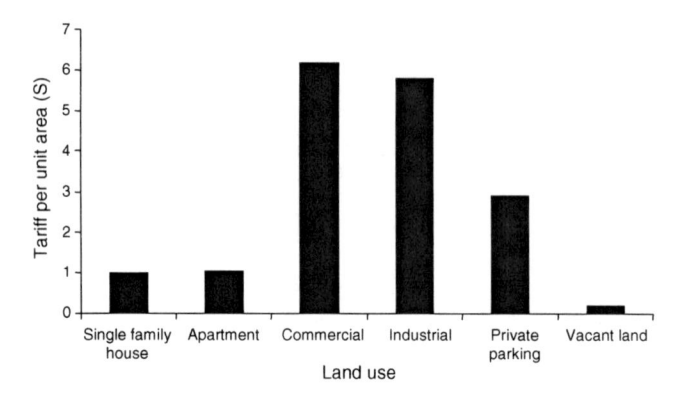

Figure 12.2 Typical monthly stormwater bills per plot for various land use classifications in Santa Cruz, California (Null 1995).

stormwater discharges from their land via the adoption of source control technologies (see Chapter 7). With the development of remote sensing combined with Geographical information systems (GIS) technology (see Chapter 5), it becomes feasible to enable an accurate assessment of the areas of impermeability.

A stormwater utility fee may provide a more stable source of revenue for local government and public utilities to invest in stormwater management programmes, which is not subject to annual budget considerations such as the imposition of tax cuts and the possibility of appropriation for other public investments. Thus, funds generated from a utility fee may be directed specifically to those areas where investments are needed and, due to the stability of the revenue, it is easier to plan for these investments. However, the charge may be overly complex which may cause difficulties in interpretation and implementation as well as additional administrational overheads (Livingston et al. 1997).

12.6 REFERENCES

Heaney, J. P., Sample, D. and Wright, L. (2002) *Costs of Urban Stormwater Control.* USEPA Report EPA/600/R-02/021, Urban Watershed Management Branch, Edison, NJ, USA, January.

Lindsey, G. (1990) Charges for urban runoff: issues in implementation. *Water Resources Bulletin* **26**(1), American Water Resources Association. February.

Manase, G., Ndamba, J. and Fawcett, B. (2002) Cost recovery for sanitation services. *28th WEDC International Conference*, Kolkata (Calcutta), India, 149–152.

Merz, B., Kreibich, H., Thieken, A. and Schmidtke, R. (2004) Estimation uncertainty of direct monetary flood damage to buildings. *Natural Hazards and Earth System Sciences* **4**, European Geosciences Union, 153–163.

Null, R. (1995) User fees – the key to managing stormwater costs. *Public Works.* Public Works Journal Corporation, Ridgewood, USA.

Paul, S. (1998) *Making Voice Work: The Report Card on Bangalore's Public Services.* World Bank Development Research Group. World Bank, Washington, DC.

Souza, C. (2001) Participatory budgeting in Brazilian cities: limits and possibilities in building democratic institutions. *Environment and Urbanization* **13**(1), International Institute for Environment and Development, London, 159–184.

Tang, J. C. S, Vongvisessomjai, S. and Sahasakmontri, K. (1989) *Estimation of Flood Damage for Bangkok and Vicinity*. Research Paper Series: IE:M No. 235. Asian Institute of Technology.

Tucci, C. E. M. (2001) Urban drainage management. In: *Urban Drainage in the Humid Tropics* (ed. Tucci), UNESCO International Hydrology Programme (IHP-V) *Technical Documents in Hydrology* **40**(1), UNESCO, Paris 2001, 157–176.

UNCHS (1993) *Maintenance of Infrastructure and its Financing and Cost Recovery*. United Nations Centre for Human Settlements, Nairobi, Kenya.

WRc (2001) *Sewerage Rehabilitation Manual*, 4th edn Version 2. WRc Swindon, United Kingdom.

ANNEX 1
Recommended reading

This following list contains books and reports that provide more information about various issues related to urban stormwater drainage in developing countries. The list contains some particularly important texts that are referenced in the book as well as other publications that are recommended sources for further reading. More details about many of these publications and how to obtain them are available on the SANICON website on the internet at www.sanicon.net.

Chapter 1   Urbanisation and urban hydrology

Burton, G.A. and Pitt, R.E. (2002) *Stormwater effects handbook – a toolbox for watershed managers, scientists, and engineers*. Lewis Publishers, A CRC Press Company, Boca Raton London New York Washington, D.C.

Ellis, JB. Chocat, B. Fujita, S. Marsalek, J. Rauch, W. (2004) *Urban drainage: a multilingual glossary*. IWA Publishing, London, UK.

Campbell, N., D'Arcy, B. Frost, A., Novotny, V. and Sansom A. (eds) (2005) *Diffuse pollution – an introduction to the problems and solutions*. IWA Publishing, London, UK.

Casale, R. and Margottini, C. (Eds) (1999) *Floods and landslides – integrated risk assessment*. Springer.

Maksimovic, C. and Tejada-Guibert, J.A. (eds) (2001) *Frontiers in urban water management*. IWA Publishing, London, UK.

Novotny, V. (2002) *Water quality: diffuse pollution and watershed management* 2nd Edition, John Wiley & Sons, Chichester, UK.

Walesh, S.G. (1989) *Urban surface water management*. John Wiley & Sons, Chichester, UK.

© 2005 IWA Publishing. *Urban Stormwater Management in Developing Countries* by Jonathan Parkinson and Ole Mark. ISBN: 1843390574. Published by IWA Publishing, London, UK.

Annex 1: Recommended reading

Chapter 2 Impacts of flooding on society

Cairncross, S. and Ouano, E.A.R. (1991) *Surface water drainage for low-income communities.* WHO/UNEP, World Health Organization, Geneva, Switzerland.
Cairncross, S. and Feacham, R.G. (1993) *Environmental health engineering in the tropics: an introductory text.* 2nd edition, John Wiley and Sons Ltd. Chichester, UK.
Few, R. (2003) Flooding, vulnerability and coping strategies: local responses to a global threat. *Progress in Development Studies* 3(1), 43–58. Hodder Arnold, London, UK.
Hardoy, J.E. Cairncross, S. and Satterthwaite D. (eds.) *The poor die young: housing and health in Third World cities.* Earthscan, London, UK.
Hardoy, J.E. Mitlin D. and Satterthwaite, D. (2001) *Environmental problems in an urbanizing world: finding solutions for cities in Africa, Asia And Latin America.* Earthscan, London, UK.
Menne, B., Pond, K., Noji, E.K. and Bertolini, R. (2000) *Floods and public health consequences, prevention and control measures.* United Nations.
Metzger, M.E. (2004) *Managing mosquitoes in stormwater treatment devices.* Publication 8125. Division of Agriculture and Natural Resources, University of California. Vector-Borne Disease Section, California Department of Health Services, USA.
WHO (2002) *Flooding: health effects and preventive measures.* WHO Fact sheet 05/02. Copenhagen and Rome, 13 September 2002.
McGranahan, G. (1991) *Environmental problems and the urban household in Third World countries.* Stockholm Environment Institute, Sweden.

Chapter 3 Integrated framework for stormwater management

Ferguson, B.K. (1998) *Introduction to stormwater: concept, purpose, design.* John Wiley and Sons, Chichester, UK.
Global Water Partnership, Technical Advisory Committee (TAC) (2000) *Integrated Water Resources Management*, TAC Background Papers, No. 4.
Global Water Partnership, Technical Advisory Committee (TAC) (2003) *The Dublin principles for water as reflected in a comparative assessment of institutional and legal arrangements of integrated water resources management*, TAC Background Papers, No. 3.

Chapter 4 Policies and institutional frameworks

Global Water Partnership, Technical Advisory Committee (TAC) (2003) *Effective water governance*, TAC Background Papers, No. 7.
Gopalakrishnan, C., Tortajada, C. Biswas, Asit K. (eds) (2005) *Water institutions: policies, performance and prospects.* Springer, London, UK.
Kuks, S., Bressers, H. (eds) (2004) *Integrated governance and water basin management – conditions for regime change and sustainability.* Environment & Policy Series, Vol. 41. Springer, London, UK.
Wallace, J., Wouters, P. and Pazvakavambwa, S. (2005) *Hydrology and water law – bridging the gap.* IWA Publishing, London, UK.

Chapter 5 Planning for urban stormwater management

Bannister, A., Raymond, S. and Baker R. (1998) *Surveying.* Pearson Longman, London, UK.
Hamdi, N. and Goethert R. (1997) *Action planning for cities: a guide to community practice.* John Wiley & Son Ltd, Chichester, UK.

Paulsson, B. (1992) *Urban applications of satellite remote sensing and GIS analysis.* UMP Discussion Paper No. 9, Urban Management Programme, World Bank, Washington, D.C., USA.

Pickering, D., Park, J.M. and Bannister. (1993) *Utility mapping and record keeping for infrastructure.* UMP Discussion Paper No. 10, Urban Management Programme, World Bank, Washington, D.C., USA.

Schultz, G.A, and Engman, E.T. (eds) (2000) *Remote sensing in hydrology and water management.* Springer, UK.

Tayler, K., Parkinson, J. and Colin J. (2003) *Urban sanitation: a guide to strategic planning.* Intermediate Technology Publications.

Uren, J. and Price W.F. (1994) *Surveying for engineers.* Palgrave Macmillan, London, UK.

Chapter 6 Approaches to urban drainage system design

Ackers, J.C., Butler, D. and May, R.W.P. (1996) *Design of sewers to control sediment problems.* CIRIA Report R141. Construction Industry Research and Information Association, London, UK.

Akan, A.O. and Houghtalen, R.J. (2003) *Urban hydrology, hydraulics, and stormwater quality: engineering applications and computer modeling.* John Wiley & Sons, Chichester, UK.

Butler, D. and Davies, J. (2000) *Urban drainage.* E & FN Spoon, London, UK.

Novotny, V., Imhoff, K.R., Olthof, M. and Krenkel, P.A. (1989) *Karl Imhoff's handbook of urban drainage and wastewater disposal.* John Wiley & Sons, Chichester, UK.

Shaw, E. (1994). *Hydrology in practice.* Chapman & Hall, London, UK.

Tucci, C.E.M. (2001) *Urban drainage in the humid tropics.* UNESCO International Hydrology Programme (IHP-V) Technical Documents in Hydrology, No. 40, Vol. 1, UNESCO, Paris 2001.

US-EPA (1997) *Guidance manual for implementing municipal storm water management programs.* Volume I – Planning and Administration. U.S. Environmental Protection Agency. Office of Wastewater Management, Municipal Support Division, Municipal Technology Branch, Washington D.C., USA.

WEF (1999) *Prevention and control of sewer system overflows.* 2nd Edition. Water Environment Federation (WEF). MOP FD-17.

WEF/ASCE, (1992) *Design and construction of urban stormwater management practices.* Water Environment Federation, Alexandria, VA, American Society of Civil Engineers, Washington, D.C.

Wilson, E.M. (1990) *Engineering hydrology.* 4th edition Palgrave Macmillan. London, UK.

Chapter 7 Ecological approaches to urban drainage system design

Girling, C., Kellett, R., Rochefort, J. and Roe, C. (2000) *Green neighbourhoods – planning and design guidelines for air, water and urban forest quality.* Center for Housing Innovation, University of Oregon, USA.

Heaney, J.P., Pitt, R., Field, R. and Chi-Yuan Fan (1999) *Innovative urban wet-weather flow management systems.* National risk management research laboratory office of research and development. U.S. Environmental Protection Agency. Cincinnatl, OH 45268. EPA/600/R-99/029.

Leonard, O.J. and Sherriff, J.D.F. (1992) *Scope for control of urban runoff.* Vol 3: Guidelines. CIRIA Report R124. Construction Industry Research and Information Association (CIRIA), London, UK.

Martin, P et al (2001) *Sustainable urban drainage systems – best practice manual for England, Scotland, Wales and Northern Ireland.* CIRIA Research Project 523. Construction Industry Research Information Association, London.
Mitchell, G., Mein, R. and McMahon, T. (1999) *The reuse potential of urban stormwater and wastewater.* Industry Report 99/14. Co-operative Research Centre for Catchment Hydrology. Monash University, Australia.
Schueler, T.R. (1987). *Controlling urban runoff: a practical manual for planning and designing urban BMPs.* Metropolitan Washington Council of Governments, USA.
Wilson, S., Bray, R. and Cooper, P. (2004) *Sustainable drainage systems. Hydraulic, structural and water quality advice.* CIRIA Research Project Report 609. Construction Industry Research Information Association, London, UK.
Wong, T., Breen, P., Somes, N. and Lloyd, S. (1998) *Managing urban stormwater using constructed wetlands.* Co-operative Research Centre for Catchment Hydrology. Industry Report 98/7. Monash University, Australia.
Wong, T., Breen, P. Somes, N. and Lloyd, S. (2000) *Water sensitive road design – design options for improving stormwater quality of road runoff.* Technical Report 00/1, Co-operative Research Centre for Catchment Hydrology, Monash University, Australia.

Chapter 8   Applications of computer models

Abbot, M.B., Basco, D.R. (1989) *Computational fluid dynamics, an introduction for engineers.* Longman Scientific & Technical, United Kingdom.
Adams, B.J. and Papa, F. (2000) *Urban stormwater management planning with analytical probabilistic models.* John Wiley & Sons, Chichester, UK.
Cunge, J.A., Holly. F.M., Verwey, A. (1980) *Practical aspects of computational river hydraulics*, Pitman Publishing Limited, London.
Mantz, P.A. (2004) *Visual hydrology – a primer for interactive computing.* IWA Publishing, London, UK.
Osman Akan, A. and Houghtalen, R.J. (2003) *Urban hydrology, hydraulics, and stormwater quality: engineering applications and computer modeling.* John Wiley & Sons, Chichester, UK.
Makropoulos, C., Butler, D. and Maksimovic, C. (1999) GIS supported evaluation of source control applicability in urban areas. *Water Science and Technology*, Vol 39 No 9, IWA Publishing, London, UK. pp. 243–252.
Mark, O. and Hosner, M. (2002) *Urban drainage modelling – a collection of experiences from the past decade.* Asian Institute of Technology, Bangkok, Thailand.
Seppelt R. (2003) *Computer-based environmental management.* John Wiley & Sons, Chichester, UK.
Soil Conservation Service, (1972) *SCS National Engineering Handbook, Section 4 Hydrology*, NRCS, Washington, USA.
Westervelt, J. (2001) *Simulation modeling for watershed management.* Springer, London, UK.
Zoppou, C. (1999) *Review of storm water models.* CSIRO Technical Report 52/99, CSIRO Land and Water, Canberra, Australia.

Chapter 9   Operational performance and maintenance

Ali, M., Cotton A. and Westlake, K. (1999) *Down to earth: solid waste disposal for low-income countries.* WEDC, Loughborough, United Kingdom.
Allison, R., Walker, T.A., Chiew, F.H.S., O'Neill, I.C. and McMahon, T. (1998) *From roads to rivers – gross pollutant removal from urban waterways.* Cooperative Research Centre for Catchment Hydrology. Report 98/6. Monash University, Australia.

Allison, R., Chiew, F. and McMahon, T. (1997) *Stormwater gross pollutants.* Co-operative Research Centre for Catchment Hydrology. Industry Report 97/11. Monash University, Australia.

Cointreau-Levine, S. (1982) *Environmental management of urban solid wastes in developing countries: a project guide.* Urban Development Technical Paper Number 5, World Bank, Washington DC.

Environmental Resources Management (ERM) (2000) *Strategic Planning Guide for Municipal Solid Waste Management.* CD-ROM prepared for the World Bank, SDC and DFID by Waste-Aware, London, UK.

Fenner, R.A. (2000) Approaches to sewer maintenance: a review. *Urban Water*, Volume 2, Issue 4, pp. 343–356.

Kolsky, P.J. (1998) Surface water drainage – How evaluation can improve performance. WELL Technical Briefs, *Waterlines* 17(1) ITDG Publishing, UK, pp. 15–18.

Kolsky, P.J. (1998) *Storm drainage: an engineering guide to the low-cost evaluation of system performance.* ITDG Publishing, London, UK.

IETC (1996) *International source book on environmentally sound technologies for municipal solid waste management.* IETC Technical Publications Series, UNEP. Japan.

Marais, M. and Armitage, N. (2002) *The measurement and reduction of urban litter entering stormwater drainage systems.* Water Research Commission Report, Pretoria, South Africa.

Read, G.F. and Vickridge, I. (1996) *Sewers: repair and renovation.* Butterworth-Heinemann.

Schübeler, P., Christen J. and Wehrle, K. (1996) *Conceptual framework for municipal solid waste management in low-income countries.* Urban Management Programme Working Paper No. 9, World Bank, Washington DC, USA.

US-EPA (2004) *Sewer sediment and control: a management practices reference guide.* January 2004 EPA 600/R-04/059. United States Environmental Protection Agency, USA.

WEF (1994) *Existing sewer evaluation and rehabilitation* – MOP FD-6. Water Environment Federation, USA.

World Health Organization (2000) *Tools for assessing the O&M status of water supply and sanitation in developing countries.* Geneva, Switzerland.

Chapter 10 Flood mitigation and response strategies

Arambepola, N.M.S.I. (2001) *Guidelines on reduction of impacts on floods.* Asian Disaster Preparedness Center, Bangkok, Thailand.

Correia, F.N., Saraiva, G., Costa, C.B., Ramos, I., Bernardo, F., Antao, P. and Rego, F. (1996) *Innovative approaches to comprehensive floodplain management: a framework for participatory valuation and decision making in urban developing areas.* Technical Annex 12, Report to European Commission, Flood Hazard Research Centre, Middlesex University, UK.

FEMA. (1998) *Homeowner's guide to retrofitting: six ways to protect your house from flooding.* FEMA No. 312. Federal Emergency Management Agency, Denver, USA.

Kreimer, A. and M. Munasinghe (eds) (1992) *Managing Natural Disasters and the Environment.* World Bank, Washington DC, USA.

Lancaster, J.W., Preene, M. and Marshall C.T. (2004) *Development and flood risk – guidance for the construction industry.* CIRIA Report R624. Construction Industry Research and Information Association, London, UK.

Chapter 11 Participation and partnerships

Abbott, J. (1996) *Sharing the city: community participation in urban management.* Earthscan, London, UK.

Affeltranger, B. (2001) *Public participation in the design of local strategies for flood mitigation and control.* Technical Documents in Hydrology No. 48. International Hydrological Programme, UNESCO, Paris, France.

Carley, M., Jenkins, P. and Smith (eds) (2001) *Urban development and civil society: the role of communities in sustainable cities* Earthscan, London, UK.

Plummer, J. (1999) *Municipalities and community participation – a source book for capacity building,* Earthscan, London, UK.

UNEP-IETC (2004) *Environmental management and community participation: enhancing local programmes.* UNEP-IETC Urban Environment Series. UNEP-International Environment Technology Centre (IETC), Japan.

UNCHS (1986) *Community participation and low-cost drainage – a training module.* United Nations Centre for Human Settlements, Nairobi, Kenya.

Chapter 12   Economics and financing

El Daher, S. (2000) *Specialized financial intermediaries for local governments: a market-based tool for local infrastructure finance.* World Bank Urban Sector – Infrastructure Notes. No. FM-8d. Washington D.C., USA.

Davey, K.J. (1993) *Elements of urban management.* UNDP/UNCHS/World Bank Urban Management Programme, Urban Management and Municipal Finance, Washington DC, USA

Green, C.H. (2003) *Handbook of water economics: principles and practice.* John Wiley and Sons, Chichester, UK.

WEF (2004) *Financing and charges for wastewater systems* – MOP 27. Water Environment Federation (WEF), USA.

WHO (1994) *Financial management of water supply and sanitation – a handbook.* World Health Organization, Geneva, Switzerland.

ANNEX 2
List of contributors

Dr Jonathan Parkinson
hydrophil
Lerchenfeldergürtel 43 Top 6/3
1160 Vienna, Austria
Email: jonathan.parkinson@hydrophil.at

Dr Ole Mark
DHI – Institute for Water and Environment
DK-2970 Hørsholm, Denmark
Email: Ole.Mark@dhi.dk

CHAPTER 1
Box 1.3 Complexities of flood management in Cape Verde
António Advino Sabino
National Institute of Water Resources
C.P. 534 – Praia, Republic of Cape Verde
Email: procave@cvtelecom.cv

Experiences from Mumbai and Hyderabad, India
Dr Kapil Gupta
Indian Institute of Technology, Bombay
Powai, Mumbai 400 076, India
Email: kgupta@civil.iitb.ac.in

CHAPTER 2
Box 2.1 Perceptions of flooding and drainage interventions in Indore, India
Simon Lewin
London School of Hygiene & Tropical Medicine
London WC1E 7HT, UK
Email: simon.lewin@lshtm.ac.uk

Flooding, vulnerability and coping strategies
Dr Roger Few
University of East Anglia
Norwich NR4 7TJ, UK
Email: r.few@uea.ac.uk

CHAPTER 3
3.4 Case study: "Slum networking"
Himanshu Parikh
Ahmedabad and Consulting Engineering
2-Sukhshanti, 10/A Purnakunj,
Ahmedabad
Gujarat, India
Email: hparikh@ntlworld.com

© 2005 IWA Publishing. *Urban Stormwater Management in Developing Countries* by Jonathan Parkinson and Ole Mark. ISBN: 1843390574. Published by IWA Publishing, London, UK.

Annex 2: List of contributors

Box 3.2 Development of the stormwater management master plan in Ipoh, Malaysia
Alias Hashim
Jurutera Perunding
17 Jalan Daud, Kg. Baru, 50300 KL
Malaysia
Email: jpzsta1@attglobal.net

Box 3.3 Integrated urban drainage system design in Vientiane, Lao P.D.R
Dr Jean O. Lacoursière
Kristianstad University, Sweden
Email: jean.lacoursiere@tec.hkr.se

CHAPTER 4
Box 4.2 New approaches towards urban stormwater management in Malaysia
Dr Nor Azazi Zakaria
(REDAC), University Science Malaysia,
14300 Nibong Tebal, Malaysia
Email: redac01@eng.usm.my

Box 4.4 Decentralised municipal management arrangements for urban stormwater management in Kampala
Albert Rugumayo
Council for Higher Education
P.O. Box 76, Kyambogo, Kampala,
Uganda
Email: nshe@infocom.co.ug

Box 4.3 Community-based watershed management in Santo André, Brazil
Erika de Castro
Centre for Human Settlements
1933 West Mall, 2nd Floor
Vancouver, BC V6T 1Z2, Canada
Email: decastro@interchange.ubc.ca

CHAPTER 5
Box 5.1 Urban Pollution Management (UPM) methodology
Dr Bob Crabtree
WRc plc
Swindon, SN5 8YF, UK
Email: crabtree_r@wrcplc.co.uk

5.4 Case study: Consultation and stakeholder analysis in Biratnagar, Nepal
Shashi Bhattarai
Integrated Consultants Nepal Pvt Ltd
P.O. Box 3839
Kathmandu, Nepal
Email: shashi@icon.com.np

CHAPTER 7
Box 7.3 Development of prototype stormwater infiltration technologies in Chile
Dr Bonifacio Fernández
Profesor de Ing. Hidráulica
Depto. Ing. Hidráulica y Ambiental
Pontificia Universidad Católica de Chile
Email: bfernand@ing.puc.cl

Sustainable drainage systems in the UK
Paul Shaffer
CIRIA, 174 – 180 Old Street
London EC1V 9BP, UK
Email: paul.shaffer@ciria.org

Box 7.1 Rainwater reuse in Bangalore
Chitra Vishwanath/S.Vishwanath
Rainwater Club
BEL Layout Vidyaranyapura
Bangalore 560 097, India
Email: chitravishwanath@vsnl.com

7.6 Case study: The BIOECODS project
Dr Aminuddin Ab. Ghani
REDAC, University Science Malaysia,
14300 Nibong Tebal, Malaysia
Email: redac02@eng.usm.my

CHAPTER 8
8.4 Case study: Urban stormwater drainage system in Addis Ababa, Ethiopia
Dirk Muschalla and Dr Manfred Ostrowski
Technische Universität Darmstadt
Petersenstr. 13, 64287 Darmstadt,
Germany
Email: muschalla@ihwb.tu-darmstadt.de

CHAPTER 9
Box 9.5 Operation and maintenance of drainage infrastructure in Hanoi, Vietnam
Dr Nguyen Viet Anh
CEETIA
Hanoi University of Civil Engineering
55 Giai Phong Road
Hanoi, Vietnam
Email: thnhueceetia@hn.vnn.vn

Box 9.3 Experiences in operation of retention basins from Belo Horizonte, Brazil
Dr Márcio Benedito Baptista and
Dr Nilo de Oliveira Nascimento
Escola de Engenharia da UFMG
Av. do Contorno 842, 8 andar
30110-060 – Belo Horizonte, MG, Brazil
Email: marbapt@ehr.ufmg.br
Email: niloon@ehr.ufmg.br

Box 9.4 Drain cleaning and solid waste management in Lahore, Pakistan
Youth Commission for Human Rights
122, Street No. 3
Lahore Cantonment, Pakistan
Email: info@ychr-crt.org

CHAPTER 10
Box 10.1 Non-structural flood mitigation measures for Dhaka City
Dr I.M. Faisal
North South University
Banani, Dhaka 1213, Bangladesh
Email: msdh@bdcom.com

Box 10.3 Flood hazard map distribution in Japan
Professor Masatoshi Shidawara
Aichi Institute of Technology
Yachigusa 1247, Toyota 470-0392, Japan
Email: shidawara@aitech.ac.jp

CHAPTER 11
Box 11.2 Community contracting for drainage channels construction
Dr Alphonce G. Kyessi
University of Dar es Salaam
P.O. Box 35176
Dar es Salaam, Tanzania
Email: kyessi@uclas.ac.tz

Box 11.1 Environmental management of an urban catchment in Lucknow, India
Shaleen Singhal
Water Resource Policy and Management
India Habitat Centre (IHC)
Lodi Road, New Delhi 110 003, India
Email: ssinghal@teri.res.in

Box 11.4 Role of women in the design of flood risk reduction and disaster mitigation in Jaleshwor municipality, Nepal
CARE Nepal
Krishna Galli, Patan,
P.O. Box 1661, Kathmandu, Nepal
Email: care@carenepal.org

CHAPTER 12
Box 12.2 Demand and willingness to pay for drainage and other sanitation services in Zimbabwe
Dr Gift Manase
Institute of Water and Sanitation Development (IWSD)
Mount Pleasant, Harare, Zimbabwe
Email: gmanase@iwsd.co.zw

Box 12.6 User fees – the key to managing stormwater costs
Roger Null
Kennedy/Jenks Consultants
622 Folsom Street
San Francisco, CA 94107 USA
Email: RogerNull@KennedyJenks.com

Index

accountability 174, 190–1
Acts 53, 54
acute impacts 12
Addis Ababa, Ethiopia 133–4
administrative boundaries 13, 42–3
adsorption 107
Aedes aegypti 23
aerial remote sensing 80–3
aesthetics of retention basins 97–8
Ahmedabad, India 47
airborne radar 81
amenity areas 96, 97
Analytical Hierarchy Process 75, 77
Ancylostoma duodenale 22
Anopheles spp. 23
Ascaris lumbricoides 22
assessment of improvement options 66–83
attenuation 5–6, 93–8
awareness raising 118, 156

bacterial diseases 22, 24
baffles 93
Ballerup, Denmark 135
Bangalore, India 108, 109, 191
Bangkok, Thailand 96–7, 165, 168, 197
Bangladesh flood warning system 168–9
basins *see* detention ponds; retention ponds
Belo Horizonte, Brazil 148–9
benchmarking 143
beneficiary assessment 176
Bio-Ecological Drainage System (BIOECODS) project 117–19
biodegradation 106
biological uptake 107
Biratnagar, Nepal 75–7
Bogotá, Columbia 162

boundaries 13, 42–3
Brazil
　Belo Horizonte 148–9
　Brasilia 6, 7
　Curitiba 4
　participatory budgeting 190–1, 192
　Porto Alegre 4, 6, 7, 96, 97
　Santa Lucia 148–9
　São Paulo
　　community response to floods 28–9
　　community-based watershed management 57–8
　　impervious areas/population density relationship 4
　　pedestrian pathways/drainage channels combination 92, 95
　　pollution problems 10
　　rainfall intensities 6, 7
　　steep drains 92, 94, 95
budgeting (municipal) 190–1
buildings
　on natural drainage pathways 2, 44–7
　structural protection 162–4
buried pipe systems 89–90

calibration of models 128–31
capacity building 61, 62–3
Cape Verde 15–16
catchments
　boundary definition problems 79
　IUWM approach 38, 39–43
　management levels 41
CBWM *see* community-based watershed management
channels, surface 88–9, 90–3
checkwalls 93

children 27
Chile 53, 54, 113, 114
cholera 22
chronic impacts 12
Cirebon, Indonesia 25
climate change 7–8
climatic factors 6–8
closed drainage systems 89–90
collection of information 77–8
Columbia 162
combined drainage systems 88
communication strategies, flood warning 165, 168
community contracting approach 178–82
community participation
   benefits 173
   investment scale reduction 196
   operation and maintenance 182–3
   planning and design 175–7
community-based approaches
   drain cleaning 182–3
   drain construction process 181
   project implementation 178
community-based organisations 182
community-based watershed management (CBWM) 57–8
computer models 121–39
   drainage systems design considerations 99–100, 102
   solutions evaluation 73–4, 77
conceptual models 123, 129
conduits drainage capacity 101–2
configuration of drainage systems 84–102
consensus agreements 173
constructed wetlands 106, 107, 115–16
construction management community participation 180
construction materials 196
consultation approach 76–7
contingent valuation methodology (CVM) 192–4
contractors 178
coping strategies 28, 185
costs
   flood damages estimation 196–7
   recovery types 198–200
   reduction 195–6
   stormwater management 194–7
Culex quinquefasciatus 23
culverts 89
Curitiba, Brazil 4
CVM *see* contingent valuation methodology

damage models 196–7
data acquisition 127–8
data-centred approaches 99, 100
debris 11
decentralised municipal management 59–60
decision-making participants 174
decision-support systems 75–7

default values 127
DEM *see* digital elevation models
demand for services 191–4
demographic trends 1–2
demonstration projects 116–19
dengue fever 23
Denmark, Ballerup 135
densification of population 4
depression storage capacity 5
depth-damage functions 197
design, urban hydrology considerations 98–101
design capacity costs 194
design storms 99, 100, 132–4
designation of responsibilities 52
detention ponds 93–4, 96, 97, 106
detention tanks 147
deterministic models 123
developing countries, challenges 13
development
   controls 96, 157, 159–62
   uncontrolled 56–7
Dhaka, India
   emergency distribution point 166
   flood inundation maps 136, 137
   flooding/city life disruption 19
   land disputes/drainage construction 13, 14
   litter in drainage channels 145–6
   non-structural flood mitigation measures 157
   urban flood modelling 137–8
digital elevation models (DEM) 125, 136, 137
digital terrain models 135
direct contamination 22
disasters
   management programmes 185–6
   mitigation cycle 184
   planning 184–6
   social amplification 25
disease transmission 21–4, 146, 149, 150
dissolved solids 11
dominant processes 126
Dominican Republic 48–9
drainage committees 181
drainage models *see* models
drainage systems
   blocked 146, 147
   cleaning 151–3, 182–3
   combined 88
   community-based construction 181
   conduit capacities 101–2
   configuration 84–102
   construction tasks 178, 179
   costs 194–7
   disease transmission routes 21–4
   environmental issues/economic development 2
   environmentally sustainable 104–19
   health impacts 21–4
   hydrology design considerations 98–101
   institutional challenges 13–14

Index

local community uses 34
major/minor interactions 85–7
public goods 188–90
'roads-as-drains' 44, 86–7
separate 87
steep slopes 92–3, 94, 95
stormwater runoff attenuation 93–8
surface 88–9, 90–3
taxation reform 55
underground 88–90
dry proofing 164
Dublin Statement on Water and Sustainable Development 35

ecological approaches 104–19
economic issues 188–200
  efficiency 35, 36
  environmental issues 2–4
  flooding impact 18–20
  rainwater reuse feasibility 110–11
education of public 118
elderly people 27
elephantiasis 23
emergency management participation 184–6
emergency (reactive) maintenance 144–5
emergency response strategies 164–9
EMP see environmental management plan
enabling environments 36
energy dissipation devices 92–3, 94
enforcement of policies 53
environment
  economic development 2–4
  runoff control policies 55–6
  runoff impact 10–12
  sustainability and IWRM principles 35, 36
environmental management plan (EMP) 175–7
environmentally sustainable drainage systems 104–19
erosion 102, 146
error sources 129–30
Escherichia coli 22
Ethiopia 133–4
European Water Framework Directive 138
evacuation routes 169
evaluation of improvement options 73–7
evapotranspiration 5
existing situation reviews 70–3, 132
extreme flood characteristics 9

faecal–oral transmission route 22
favelas (slums) 57
feedback 191
filariasis 23
filtration 106, 107
financial issues 55, 111, 188–200
flash flooding 5–6
flooding
  attenuation capacity reduction 5–6
  causes 8–10
  community/personal responses 27–9

coping strategies 28, 185
damage costs 196–7
definition 10
Dhaka modelling case study 137–8
economic impacts 18–20
emergency response strategies 164–9
evacuation routes 169
flash 5–6
hazard maps 165, 166
health impacts 19, 21–4
insurance 170–1
inundation maps 135–8, 168–9
livelihood impacts 24–7
mitigation cycle and stages 156–71
mitigation model case studies 133–4, 137–8
non-structural control strategies 155–71, 183–6
past experience review 158
perceptions 19, 27–9
personal vulnerability 20, 24–7
physical impacts 8–10, 18, 20
prediction 28
preparation 159–60
proofing measures 157, 159–60, 162–4
psychological effects 20
recovery strategies 160, 169–71, 185
rehabilitation 160, 169–71
response strategies 27, 29–30, 160, 164–9, 184–6
return frequency 100–1
return period 194–5
risk assessment 156–8
routing 86–7
simulation results application 135–8
social impacts 18–20
society impacts 18–30
types 8–10
urban drainage/stormwater models 124, 125
warning systems 27, 156, 157, 159–60, 165, 167–9
women 27, 174, 185–6
zoning 161
floodplains 2
floodwalls 163, 164
flukes 24
flushing of storm drains 153–4
funding sources 198–200

gender perspectives 185
general taxes 198
geo-reference 83
geographical information systems (GIS) 78, 127–8, 135–6, 162, 168–9
geographical positioning systems (GPS) 80
George Compound, Lusaka 183
GIS see geographical information systems
global urban population growth 1–2
Gokwe, Zimbabwe 193
governance scorecards 190, 191
GPS see geographical positioning systems

grassed swales 112–13
gravity drainage 101
greases 11
Greece, Thessaloniki 135
groundwater recharge 107, 109, 110, 111
growth urban population (global) 1–2
gully pots 150
Gutu, Zimbabwe 193

Haiti 82
hazard maps 165, 166
health issues 19, 21–4, 30, 146, 149, 150
heavy metals 11
helminths 22
herbicides 11
hookworm 22
human resources 62–4
Hyderabad, India 2, 14, 15
hydraulic gradient 101
hydrocarbons 11
hydrological boundaries 13, 42
hydrology
 drainage design considerations 98–101
 urbanisation impact 1–6

IDF *see* intensity–duration–frequency
Ikonos 81, 82
illness *see* health issues
image archives 82–3
impervious areas and economic development 3–6
improvement options
 evaluation and comparison 73–7
 planning and assessment 66–83
inclusivity 174
India
 Ahmedabad 47
 Bangalore 108, 109, 191
 Baroda 47
 Delhi 15, 110
 Hyderabad 2, 14, 15
 Indore 28, 45, 46, 47, 87
 Lucknow 175–7
 Mumbai 7, 47
 Warangal 8
indicators 142–3
Indonesia 25
Indore, India 28, 45, 46, 47, 87
infiltration 111–12, 113
 basins 107
 BIOECODS project 117–19
 capacity reduction 4
 impervious areas increase 4
 pollutants removal 106
 runoff control/groundwater recharge 107
 trenches 107, 113, 115
 wells 114
informal settlements
 cities population proportion 2
 cost recovery complications 192

institutional challenges 13
land use policies 57–8
information collection and management 77–8
ingestion of pathogens 22
inlets 90
inorganic compounds 11
institutional frameworks 36, 51–64
institutions
 development 60–4
 flood control strategies co-ordination 157
 runoff control policies 54–5
 urbanisation challenges 13–16
insurance 170–1
integrated approaches
 flood warning systems 168–9
 modelling 138–9
 stormwater management 33–49
integrated urban water management (IUWM) 33–49
 catchment planning 38, 39–43
 infrastructure/services 38, 43–7
 main components 38–9
 partnerships 38, 47–9
 stakeholder participation 38, 47–9
integrated water resources management (IWRM) 33–49
 key elements 35–9
 principles 35
intensity–duration–frequency (IDF) curves 99
inundation definition 10
inundation maps 134–8, 168–9
investments reduction 195–6
Ipoh, Malaysia 37–8
Iran 153
IUWM *see* integrated urban water management
IWRM *see* integrated water resources management

Jakarta, Indonesia 25
Jaleshwor, Nepal 185–6
Japan, flood hazard maps 166
job performance 62, 63

Kampala, Uganda 60

Lahore, Pakistan 151
land disputes 13, 14
land use
 classification system 199–200
 controls 157, 159–62
 policies 56–8
 user's fees 199–200
land-based topographical surveys 80
Landsat 80–1, 82
landslides 26, 162
learning/job performance relationship 62, 63
lease contracts 182
leptospirosis 24
LIDAR 81
literature values 127

litter 145–6
livelihood flooding impact 24–7
livelihood strategies
local authorities 58–9
local community knowledge 72
local contractors 178
local governments 58–9
local stakeholders 173, 178–82
local taxation 198
low-cost structural building adaptations 164
low-income groups
  participation benefits 173
  vulnerability 24–7
  willingness to pay 192–3
Lusaka, Zambia 183

maintenance 140–54
  emergency (reactive) 144–5
  existing systems 72
  participation 182–3
  periodic 144
  rehabilitation work 144
  routine 144
  solid waste control problems 149–54
  strategies 143–5
major drainage systems 85–7
major flood characteristics 9
malaria 23–4
Malaysia
  BIOECODS project 117–19
  Ipoh stormwater management master plan 37–8
  runoff control policies 55, 56
  swales 114
  Universiti Sains Malaysia 114
Manning numbers 130
maps
  flood hazard 165, 166
  flood inundation 135–8, 168–9
  flood-zoning 161
  spatial 79–83
master plans 68
mathematical models 123, 129, 130
MCDA *see* multi-criteria decision-making analysis
measurements of operational performance 141–3
mega-cities 2
melioidosis 22
minor drainage systems 85–7
minor flood characteristics 9
models
  application 132–8
  building 127–8
  calibration 128–31
  computer 121–39
  data acquisition 127–8
  design storms 132–4
  dominant processes 126

drainage design considerations 99–100, 102
  error sources 129–30
  flood simulation 135–8
  integrated 138–9
  modelling procedures 125–38
  objectives 126
  performance analysis 132–4
  planning and preparation 126–7
  procedure steps 125–6
  selection 127
  sensitivity analysis 128
  solutions evaluation 73–4, 77
  types 123
  uses 124–5
  validation 128–32
moderate flood characteristics 9
mosquitoes
  closed systems 90
  constructed wetlands 115–16
  construction sites 30
  disease transmission 22–4, 150
  open channels 91
multi-criteria decision-making analysis (MCDA) 75–7
multi-spectral data 81
municipal budgeting 190–1
municipal workers' training 149, 150

national disaster management systems 13
natural attenuation capacity loss 5–6
natural drainage pathways 2, 44–7
natural habitats 106
Necator americanus 22
nematodes 22
Nepal 75–7, 185–6
nitrogen 11
non-structural flood mitigation strategies 155–71, 183–6

O&M *see* operation and maintenance
off-line attenuation ponds 94, 96
off-site technologies 105–7
oils 11
on-line attenuation ponds 94, 96
on-site technologies 105–7
open channels 88, 90–3
operation and maintenance (O&M) 140–54
operational performance 140–54
  existing systems improvement 72
  measurements 141–3
  policy development influence 67
  solid waste 145–9
  strategies 143–5
operational sustainability 140–3
Orbview 3 satellite 81, 82
organisation strengthening 60–4
overflows 55, 88
overland flow 99–100
oxygen demanding material 11

Pakistan 151
  Dhaka
    emergency distribution point 166
    flood inundation maps 136, 137
    flooding/city life disruption 19
    land disputes/drainage construction 13, 14
    litter in drainage channels 145–6
    non-structural flood mitigation measures 157
    urban flood modelling 137–8

pan-sharpened data 82
panochromatic imagery 81
parasitic helminth infections 22
parking areas 3–4
participation 172–7
  extent 173
  forms 172–4
  investment scale reduction 196
  IUWM approach 38, 39, 47–9
  key principles 173–4
  non-structural flood control strategies 183–6
  operation and maintenance 182–3
  participants 174
  planning and design 174–7
  potential benefits 172–4
  women's role 185–6
participatory assessments 75
participatory budgeting 190–1, 192
Participatory Rapid Appraisal (PRA) 176, 185
partnerships 38, 47–9, 171, 177–82
past experience review 158
pathogens 21–4
paving 4
pedestrian pathways/drainage channels combination 92, 95
perceptions
  flooding 19, 27–9
  local government effectiveness 190
  planning process 69
performance
  analysis models 132–4
  evaluation 140–3
  indictors 142
  operational 140–54
periodic maintenance 144
permeable surfaces 106, 107, 111, 112, 113
permit fees 198–9
pesticides 11
phosphorus 11
photography 81–3
physically-based models 122, 123, 129
piles 163
piped systems 89–90
pits 112
planning
  decision-support systems 75–7
  definition 66–8
  existing situation assessment 70–3

improvement options 66–83
information collection and management 77–8
institutional challenges 13–14
modelling procedures 126–7
problem analyses 71–3
process 68–70
regulations disregard 14
solutions formulation 71–3
spatial mapping 79–83
Plasmodium 23
plug-flow-pollution models 124
policies 51–64
  development 68
  formulation 51–4
  implementation implications 53–4
  institutional development 60–4
  institutional frameworks 58–60
  land use 56–8
  making procedures 53
  organisational strengthening 60–4
  principles of 52
  runoff control 54–6
  types of 51–2
pollution
  constructed wetlands 115
  on-site/off-site technologies 105–7
  protection policies 55
  removal mechanisms 106, 107
  runoff control policies 55
  runoff impacts 10–12
  source control 105
  swales 112–14
  urban drainage/stormwater models 124–5
polythene 146
population growth 1–2, 4
Port Au Prince, Haiti 82
Porto Alegre, Brazil 4, 6, 7, 96, 97
poverty and vulnerability 24–7
PRA *see* Participatory Rapid Appraisal
practical demonstration projects 116–19
pre-disaster planning 164
prediction of flood situations 28
prevention planning 42, 67
private developers 14, 15
private goods 189
private sector 182, 196
problem analyses 71–3
process indicators 142–3
programme planning 68
projects
  partnerships in implementation 177–82
  planning stage definition 68
proofing, flood 157, 159–60, 162–4
protozoa 23
psychological effects of flooding 20
public awareness 118, 156
public goods 188–90
public information sources 168
public participation 175, 184–6

Index

public perception of flooding 19, 27–9
pumping stations 101

qualitative assessments 72–5
quantitative assessments 72–4
questionnaires 175
QuickBird 81, 82

rainfall
   drainage systems design considerations 99–101
   global distribution 6
   intensity and duration 6–7, 99
rainfall-runoff simulation models 100
rainwater reuse 107, 108–11
Rational Method 99
rats 24
reactive maintenance 144–5
real-time systems 167–9
recovery issues 160, 169–71, 185, 191–2, 198–200
recreational areas 97
recurrence intervals of flooding 98, 100–1
reforestation 133, 134
refuse 145–6
rehabilitation 144, 160, 169–71
relocation programmes 19, 57–8, 161–2
remedial solutions 67
remote sensing 80–3
report cards 190, 191
resettlement programmes 19, 57–8, 161–2
resources
   existing situation reviews 70–3
   policy implementation 53
respiratory diseases 21
responsibilities designation 52
retention ponds 93–4, 96–8, 106, 148–9
retention tanks 147
return frequency of flooding 98, 100–1
reuse of rainwater 107, 108–11
revenue generation 54, 198–200
risk management 156–8
risk reduction runoff control policies 55
roads
   environmental issues 3–4
   groundwater replenishment from runoff 110
   'roads-as-drains' 44, 86–7
   runoff and pollution 12
roofs 4, 108, 109
roundworms 22
routine maintenance 144
runoff
   attenuation 93–8
   climatic factors 6–8
   drainage systems design considerations 99–100
   environmental impacts 10–12
   groundwater replenishment from road runoff 110
   impervious areas 4
   land use classification system 199–200
   pollutant removal mechanisms 106, 107
runoff control
   constructed wetlands 115
   on-site/off-site technologies 105–7
   options 102
   policies 54–6
   source control 105
   swales 112–14

Saint Venant equations 123, 125
SAMBA model 124
sanitation 2–3, 149, 150
Santa Cruz, California 199–200
Santa Cruz, India 7
Santa Lucia, Brazil 148–9
Santo André, Brazil 57–8
Santo Domingo, Dominican Republic 49
São Paulo, Brazil
   community response to floods 28–9
   community-based watershed management 57–8
   impervious areas/population density relationship 4
   pedestrian pathways/drainage channels combination 92, 95
   pollution problems 10
   rainfall intensities 6, 7
   steep drains 92, 94, 95
satellite imagery 80–3
schistomiasis 24
scorecards 190, 191
sediment traps 150
sediments and sedimentation
   environmental impacts of pollution 11
   operational performance impact 146–7
   pollutant removal processes 107
   pollution control 106
   retention basins 96–7, 148–9
self-cleansing velocity 89–90, 101–2
sensitivity analysis, models 128
separate drainage systems 87
Serbia 89
services
   charges 54, 199–200
   demand and willingness to pay 191–4
SIMPOL model 74, 124
simulations 131
skills, policy implementation 53
slums 44–7, 87
small-format aerial photographs 81
snails 24
soakaways 107, 110, 111, 112
social amplification 25
social equity 35, 36
social impacts of flooding 18–20
social interaction encouragement 182
society and flooding impact 18–30
soil erosion 146

Index

solid waste
  closed drainage systems 90
  control problems 149–54
  environmental issues 2–3
  investment scale reduction 196
  management 2–3, 157, 196
  operational performance impact 145–9
  types and constituents 146
solids size distribution in open drains 146, 147
solutions
  evaluation/comparison 73–7
  formulation and problem analyses 71–3
source control 105, 196
spatial data mapping 79–83
spatially distributed data 78
SPOT 80–1, 82
SQIRTS see Stormwater Quality Improvement and Reuse Treatment Scheme
stakeholders 48, 52, 58–9, 76–7
  see also participation; partnerships
steep slopes 92–3, 94, 95
step drains 93
stilts 163
storage of stormwater runoff 93–8
storm drain flushing 153–4
Stormwater Quality Improvement and Reuse Treatment Scheme (SQIRTS) 116
straining, pollutant removal processes 107
street sweeping 150
structural protection measures 162–4
surface drainage systems 88–9, 90–3
surface storage capacity 5
surfaces, flood control models 125
survey techniques 79–83
suspended solids 11
sustainability
  environmentally sustainable drainage systems 104–19
  IWRM principles 35, 36
  operational 140–3
swales 106, 107, 112–15, 117–19
Sydney, Australia 116
synthetic design storms 99, 132

Tanzania 179, 180
taxation 55, 198
technical capacity 53
Tehran, Iran 153
temperature 7
Thailand, Bangkok 96–7, 165, 168, 197
Thessaloniki, Greece 135
topographical surveys 79–83
total overflow volume 55
training issues 61, 62–4, 149, 150
transect walks 176, 177

transmission of disease 21–4, 146, 149, 150
transparency 174
trapezoidal channels 91, 92
trapping structures 90, 150
trash 11
Trichuris trichiura 22

Uganda 60
uncontrolled development land use policies 56–7
underground drainage systems 88–90
Universiti Sains Malaysia 114
Urban Pollution Management (UPM) 74, 138
urban yellow fever 23
urbanisation issues 1–16
users' fees 199–200
utility fees 199–200

validation, models 128–32
vegetation 106, 107, 112–14, 115
Vientiane, Lao PDR 34, 39–40
Vietnam 92, 152–3
viral diseases 23
vulnerability, personal to flooding 20, 24–7

warning systems 27, 156, 157, 159–60, 165, 167–9
waste traps 90
'water for livelihoods' 35, 36
water quality 55, 96–7
water rates bills 54
'water as a resource' 35, 36
water resources 2–3
water temperatures 11
water-related/waterborne diseases 21–4
Weil's disease 24
wet ponds see retention basins
wet proofing 164
wetlands 106, 107, 115–16
whipworm 22
willingness to pay 190, 191–4
wind 7
women 27, 174, 185–6
worm infections 22, 23

YCHR see Youth Commission for Human Rights
yearly overflow volume 55
yellow fever 23
Youth Commission for Human Rights (YCHR) 151

Zambia 183
Zimbabwe 192, 193
zoning 161